SOILS

SOILS

THEIR NATURE, CLASSES, DISTRIBUTION, USES, AND CARE

by

J. Sullivan Gibson

and

James W. Batten

University of Alabama Press
University, Alabama

ABOUT THE AUTHORS

Dr. J. Sullivan Gibson was for many years Associate Professor of Geography at the University of North Carolina at Chapel Hill. From 1966 to 1968 he was visiting Professor of Geography at Indiana State University. His co-author, Dr. James W. Batten, is Professor of Research—with emphasis on inter-disciplinary studies pertaining to earth sciences—and Chairman of the Department of Secondary Education at East Carolina University, Greenville, North Carolina.

To a heritage that taught us to use and revere the soil,
To the masters who showed us the intricate wonders of its
life-giving qualities,
To our students who inspired us to disseminate the knowledge
of soils,
This book is dedicated.

Foreword

The study of soil as a topical field within geography has suffered from neglect for the past decade or so, during which geographers in large numbers have turned their attention to urban problems. Yet no element of man's habitat illustrates more clearly some of the underlying problems of any kind of geographic study. No two points on the face of the earth have identical soil conditions, and yet to attempt to describe every minute bit of soil would be undesirable even if it were possible. The soil specialists were among the earliest geographers to recognize the need for classifying the phenomena of the earth's surface at different degrees of generalization. Even the phase of a soil type, which might be described as occupying a part of a certain farmer's field, would nevertheless still be a generalization, based on the recognition of significant similarities and in spite of minute variations. But when the soil specialist looks at the face of the earth more broadly, the details used in the definition of soil types are no longer visible. A greater degree of generalization is needed. By looking at the world as a whole, or at major parts of it, major soil groups can be identified.

How can different degrees of generalization be used systematically to build a useful classification of soils? At one time—in the early 1920's—general soil categories for use at the county level were recognized by trying to generalize the soil types. Then the soil specialists developed a new approach. They defined repeated associations of soil types and drew lines on maps around the areas in which these associations were to be observed. For example, the *catena* is an association of soil types found characteristically at different elevations on hill slopes. The definition of soil associations was a major contribution to the method of describing the face of the earth at a degree of generalization suitable for mapping at scales intermediate between the directly observable soil types and the continental-scale soil groups.

In 1949 Charles E. Kellogg and others used a whole issue of the periodical *Soil Science* to present a new classification of soils at different degrees of generalization. The new terminology that was suggested proved very helpful to soil specialists, but it was too novel and too complicated to be easily adopted by nonspecialists.

This book by J. Sullivan Gibson and James W. Batten provides a much-needed restatement of the newest ideas about the processes of soil formation, and a simple treatment of soil classification. The book is useful as a supplementary text for courses in geography in the secondary schools and the colleges. The materials are presented in terms understandable to nonspecialists. The authors show how the soil categories are related to the soil-forming processes, and also how these categories can be used as a guide to potential uses. This is an up-to-date and readable contribution to the understanding of soil as an element of the human habitat.

PRESTON E. JAMES

Preface

Each year technological advances increase and techniques improve; each year it becomes more important for students to have opportunities to think rationally, communicate effectively, and solve problems efficiently. New courses of study that emphasize solid science content are being included in the curricula of high schools and colleges, and programs are being designed to encourage students to identify problems, propose solutions, and draw valid conclusions.

The rapid technological advances of the space age are having a definite influence on the "climate" in the classroom. Students think more abstractly than ever before and seek to be challenged by concepts rather than by rote memorization of facts. From their combined experience of more than seventy years in the classroom, the authors have concluded that opportunities for problem solving are essential to students. Among such concepts are interrelatedness and interdependence. For example, the earth is a part of the solar system, which is a minute part of the galactic system. The two systems are interrelated, and the nature of changes in each is affected by the very fact that they are. The location of man and his attitudes and activities are dependent upon the natural forces operating in the earth. Climatic and weather conditions result from atmospheric disturbances, and oceans, seas, and other bodies of water have characteristic influences on weather and climate. Mountains, valleys, and plains result from structural and gradational processes over long periods of time. Rocks, minerals, and metals found in the earth's crust result from the combination of various elements. Not to be overlooked are the *Soils*. They develop under the influences of climate, vegetation, slope and drainage, time, the nature of the parent material, and the culture. Even in the space age, we need to get out of "orbit," study the soils in place, and interpret them in the light of the natural environment of which they are a part. We must realize that despite the advances of technology which put man on the moon, the total population, over three billion people, must get food from the top six inches of the earth, and that soils must last longer than a particular culture.

All soils do not have the same utility, but man uses different soils in different ways. "Good" land for the production of foodstuffs must lie well and have good depth, for yields are dependent upon the ability of the soil to take up and use fertilizers and water. Man has done much to adapt crops to the soil and to provide various kinds of fertilizers for plant growth and development. Soils that are not good for the production of foodstuffs may be valuable in other ways. For example, Podzols in high elevations are poor for crops but they comprise excellent forest soils.

Each soil series requires skillful handling if it is to produce to its maximum potential; but no two series make the same demands. From season to season conditions of temperature and moisture change, so the farmer must change his management to produce better drainage, improve tilth, provide contours to prevent erosion, and test his soil to identify the proper kind and the correct proportion of fertilizer needed. Only by careful study of the soil, resulting in an understanding of the complexity of its nature and uses, will man be able to provide food for all the people who will inhabit the earth. The soil does not grow and increase; it cannot reproduce itself. Therefore, man should seek to improve the soil through good management and kind treatment so that subsequent generations can farm more efficiently than their fathers and grandfathers have done. Man can improve the soil now in use and even discover how more kinds of soils can be utilized more productively.

We, the authors, have loved the soil all our lives because we have tilled it, managed it, watched it produce bountiful crops, seen its production decline because of droughts or storms, appreciated its values, deplored its abuse, and bemoaned its limitations.

The obligations of the authors are many, too many to permit acknowledgements in full. Our agrarian heritage provided the first spark of enthusiasm. Through the masters with whom we studied and worked we have learned the scientific approach to soils in their proper environmental role. As great teachers we remember the late Curtis F. Marbut and the late W. Elmer Ekblaw. Three professional colleagues deserve mention: the late L. R. Schoenman, of the Tennessee Valley Authority, and Ralph J. McCracken and Eugene F. Goldston, of North Carolina State University at Raleigh, North Carolina. Students who have studied in our classes, too numerous to mention by name, provided challenge and inspiration at all times. Librarians at the University of North Carolina, East Carolina University, and Indiana State University made literature and other needed facilities available. Professor Amos O. Clark read the entire manuscript and offered valuable and constructive suggestions. Of several typists who helped with the manuscript, Jane Jacobi, Judy Marshburn, Jennifer B. Helms, Lou Anna Hardee, and Peggy Hearn should be mentioned. The Soil Conservation Service generously provided photographs which were used to illustrate the book. Finally a word of deep appreciation goes to Kay Gibson and Sara Storey Batten, the understanding wives, who provided continuous encouragement and inspiration, and exhibited great patience, while this work continued. To all these individuals, groups, and institutions we wish to express our deep appreciation.

J. SULLIVAN GIBSON
JAMES W. BATTEN

Table of Contents

List of Illustrations

SOILS

CHAPTER 1

The Nature of Soils

THE MEANING OF "SOILS": THEIR NATURAL QUALITIES

The significance of soils in providing food and industrial products is incontestable. Throughout history man has been largely dependent on the soils for nourishment. Today he is no less dependent upon them. Furthermore, the population of the world is increasing rapidly while land area and soil resources remain essentially constant, so that an understanding of soils and their place in human environment is vitally important.

Soils constitute an independent body of material in nature. The word "soil" has a number of meanings: the loose, unconsolidated material that covers the earth's crust; the surface layer of such unconsolidated material, that is, the rooting zone of plants; the broken-down, weathered pieces of rock that accumulate from rock near the surface; materials brought in from other places by natural carriers.

To the soil scientist, or pedologist, the term "soil" means all of these things and a great deal besides. The pedologist sees soil as the result of highly complex processes working slowly over long periods of time and producing varied changes in the vast array of mineral and organic materials that mantle the earth's surface. For the pedologist, soil is a new natural body of earth material, strangely different from all other natural bodies, and possessing remarkable life-giving qualities.

The composition of the earth's thin soil covering is very much akin to that of the rocks and minerals which form the major part of the earth's crust. Yet the soil body differs in many ways from the materials comprising it, and its complex nature gives rise to many thousands of distinctive soils. Soils possess the unusual ability to sustain the life of all forms of land vegetation, and when properly handled, soils retain this capacity indefinitely, without deteriorating or diminishing.

To understand soils we must view them as part of the natural environment, in relation to rocks, relief, climate, natural vegetation, and time.

Soils are part of the earth's mantle. Soils cover a large proportion of the earth's crust. There are large areas, however, where the solid earth has no soil at all. Much desert land is stripped to solid rock or strewn with broken stone. Many areas of rugged mountain are barren bedrock. More than one-tenth of the earth's land surface is covered by snowfields and glaciers. Soils have been developed only where unconsolidated earth material, called *regolith,* mantles the surface.

This layer of earth materials mantling the solid part of the earth's surface might all be classed as "soil," but only in a very general sense. Many large areas have deep soils, fully developed, others have only a film. Some broad stretches of desert have deep mantles of shifting sand. Many rugged mountain areas gather sufficient soil to nurture abundant forest growth. Even the frozen wastelands collect mantle rock underneath the ice and snow. Some glaciers have collected, along their margins, layers of broken rock and other materials that are thick enough to support trees, with the result that forest stands move slowly downslope atop the flowing ice. River floodplains gather layer after layer of sand, silt, and clay as floodwaters come and go. But not all such mantle material is "true soil"—"soil" as employed in the scientific sense by pedologists.

Much of the mantle has not developed true soil. True soil develops slowly, and, under many circumstances, the natural processes that form it never complete their work. Thus, for example, the shifting sands of deserts and of coastal dunes lack the stability necessary for complete soil formation.

Similarly, as river floodplains collect one deposit of sediment after another (with each successive inundation), the processes that might otherwise produce mature soil are interrupted. On steep slopes of land, erosion and gravity may remove loosened materials as fast as natural forces produce them from solid rock, thus preventing soil formation. Many areas of marsh and swamp never acquire true soils because they are always flooded. A considerable percentage of the earth's land surface, therefore, is without true soils.

True soils are dynamic. Soils, as understood by pedologists, are dynamic in nature. They nurture plant and animal life and are, in a sense, "living." Soils cannot reproduce themselves, of course, but they do nurture living species—plant roots, worms, bugs, and less complex life forms such as bacteria and fungi—that can continue to multiply throughout countless generations. Thus, somewhat in the manner of plants and animals, soils evolve through time, passing through stages of youth, maturity, and old age. This is one of the remarkable qualities that distinguish soils from all other natural bodies.

MATERIALS OF TRUE SOILS: MINERAL MATTER, ORGANIC MATTER, SOIL WATER, SOIL AIR

Soils comprise four types of material: (1) mineral materials derived from rocks, (2) organic materials derived from plants and animals, (3) soil water, and (4) soil air. The mineral and organic components vary greatly from one soil to another, but remain relatively constant in any particular soil. Soil water and soil air, however, vary greatly from time to time as well as from soil to soil.

The bulk of soil is mineral. Most soils are largely mineral in bulk. The *parent material* providing soil minerals is derived largely from broken and weathered rock. In most instances parent material accumulates in place from local rock, in which case it is called *residual* material. Residual parent material often develops into soils readily identified with the

Fig. 1–1. Alluvial soils on river floodplains often are highly productive if they escape flooding. This field of young corn on the banks of the Ohio River is typical of much of America's fine farm land. (Huntington silt loam; Spencer County, Indiana)

particular local rock. Thus soils derived from weathered limestone often are referred to as limestone soils.

The parent material of many soils accumulates as it is deposited by transporting agents, largely running water, wind, and glacial ice. Such *transported* materials usually are thoroughly mixed; and in many instances so altered in other ways that identification of specific parent rock is difficult or impossible. Material deposited on river floodplains, called *alluvium,* is the most widespread of transported parent materials. Glacial deposition of many different materials, designated by the general term *glacial till,* provides the bulk of parent material over large glaciated areas of North America and Eurasia. In some areas of considerable extent, such as the interior of China and the western parts of Kentucky, Tennessee, and Mississippi, wind-laid, or *eolian,* materials, called *loess,* comprise blankets many feet thick.

Organisms, dead and alive, provide vitality. Organic material of soils comes largely from leaves, stems, branches, roots, and trunks of various kinds of plants. In addition, a variety of animal forms including earthworms, bugs, insects, and small burrowing animals provide material that, though of relatively small bulk, is very important as a soil constituent. Belonging also to the organic content of soil are microscopic forms, including bacteria and fungi, which perform varied and intricate functions in soil development and behavior.

Organic material that is decomposed and assimilated into the soil is *humus*; undecomposed organic matter is *raw humus*. Some soils are so heavily charged with organic matter that they are classed as organic soils, referred to as *peat* and *muck*. They are widely distributed, particularly in swamp and marsh areas.

Fig. 1–2. Some organic soils are highly productive. Cabbages grown under irrigation are a profitable crop on this Warners muck. (Lake County, Indiana)

COURTESY OF SOIL CONSERVATION SERVICE, U.S.D.A.

An interesting feature of some organic soils is their ability
(because of their high carbon content) to burn when ignited
by either natural or artificial means. In dry seasons, when
ground water is at a minimum, "ground fires" often burn and
smolder for weeks at a time. Sometimes such fires spread into
forested lands, destroying valuable stands of timber and other
property.

Color differences in soil are frequently related to organic
materials. Light-colored desert and dune sands are very low in
organic material whereas the dark brown and black soils of
grasslands are relatively high in organic content.

Water makes soil development possible. Water is a
highly variable component of soils. Some soils are wet at all
times, requiring artificial drainage if they are to be used for
agriculture. Others, such as first bottoms of river floodplains,
are periodically wet but sufficiently drained for cultivation during low-water seasons. Most agricultural soils have adequate
water to meet vegetation requirements during a considerable
part of the year, although plants often suffer during periods of
drought. Still other soils are continuously dry, because of inadequate rainfall or excessive drainage, and will grow few
plants without irrigation.

Soil water occupies open-pore spaces in the soil. In general, fine-textured soils such as clays and loams absorb more
water; the intake, however, is slower than that of sandy soils.

Air complements water; both are essential. Soil
air is as variable as soil water. In fact, the two complement
each other to the extent that together they fill all the pore
space that is not taken by plant roots. Since air is lighter
than water, it may be crowded out of the soil as more and
more water fills pore spaces. When this situation develops the
soil is said to be *waterlogged*. Thus the function of soil air
is complicated. Although some plants, such as reeds, thrive in
permanently waterlogged soil, many plants need both air and
water in the soil most of the time. Soil air also functions in

the weathering and *decomposition* of minerals and organic materials, and in other chemical reactions that take place in
the soil. Oxidation of iron, responsible for the bright red color
of many well-drained soils in warm, humid climates, is a
prominent result of these highly important chemical processes.
The absence of air in waterlogged peat, muck, and swamp
soils results in very slow decomposition and assimilation of
organic matter.

PHYSICAL PROPERTIES OF SOILS: TEXTURE, STRUCTURE, QUANTITY AND ARRANGEMENT OF WATER, COLOR

Four physical properties, texture, structure, quantity and
arrangement of water, and color are, recognizable in all soils.
These physical properties are readily detected through the
senses of sight and touch. They are largely the "look" and
the "feel" of soils. These properties can be approached and
measured by applying the principles of physics. They are not
related to the chemical reactions or the biological relationships
of soil materials. Thus the study of the physical properties
provides the simplest method of beginning to understand soils.

Texture is the fineness or coarseness of soils.
Several texture classes of soils are readily recognized. The
casual observer distinguishes between sandy, loamy, and clayey
soils. By looking more closely, and by feeling the "grain" of
the soil, he can make other valid conclusions. Soil texture has
to do with the fineness or coarseness of soil particles. Mineral
particles, which make up the bulk of soil, vary greatly in
size. They range downward from small stones, gravel, and
sand grains, to the microscopic clay particles and sub-
microscopic particles that comprise soil *colloids*. The four principal size categories are "gravel," "sand," "silt," and "clay."
Some soils, for example sand, consist largely of particles of
approximately the same size. Most soils, however, have two or
more groups, classified by size of particles, usually with one

group dominant. Thus, in grouping soils into texture classes, the proportion of particles belonging to different size groups, as well as the particle sizes themselves, is important.

In most soils texture varies greatly from the surface downward. The subsoil usually contains more clay and other fine material than does the surface soil, although this is not always the case. In soil classification, the texture of the surface soil seems more significant than that of deeper layers. Therefore, soils are usually classified according to the texture of a six- to eight-inch thick surface layer, approximately the "plow layer." Six major texture groups are "sand," "sandy loam," "silt loam," "loam," "clay loam," and "clay." Each of these groups may be subdivided when it is useful to do so.

The choice soils, overall, are loams. Many soil qualities are closely related to texture. Since fine-textured soils have greater pore space and larger surface area than coarse-textured soils, they provide greater storage space for water and better feeding zones for plant roots. Thus, in a broad, general way, relatively fine-textured soils are more productive agriculturally than are soils with coarse texture. Too fine a texture, however, adversely affects tillage. Sands and sandy loams are more easily tilled than clays and clay loams because the tilling of the former requires less power and is hindered less by wetness.

Table 1 and Figure 1-3 depict two different ways of representing soil texture. Four soils are represented. Their textures differ widely. Particle size ranges from almost one extreme to the other. The Norfolk sandy loam, an excellent soil for tobacco, cotton, peanuts, and other crops grown widely in the Atlantic Coastal Plain, lies entirely in the sand triangle, since at each depth the sand component is more than 50%. In like manner, the Clinton silt loam, a prominent Midwestern soil, is strongly silt at all depths, and is shown in the silt triangle. The Davidson clay loam, one of the best soils in the Appalachian Piedmont, has a strong clay component in the subsoil, but sand and silt are almost as abundant as clay at

Table 1

Displayed here are the generalized mechanical analyses of four soil samples: 1. Norfolk sandy loam, 2. Clinton silt loam, 3. Davidson clay loam, and 4. Carrington silt loam. Only three size categories of soil particles are recognized here: (a) sand (generalized), (b) silt, and (c) clay.

	Horizon	Depth inches	% Sand	% Silt	% Clay	Total %
1. Norfolk sandy loam	A¹	0-2	69.4	21.4	9.1	99.9
	A²	2-8¼	67.0	21.4	11.4	98.8
	B	8½-27	58.7	19.3	22.0	100.0
	C	27-36	59.5	18.3	22.1	99.9
2. Clinton silt loam	A¹	0.12	24.2	62.1	13.0	99.3
	A²	12-20	32.1	56.1	11.6	99.8
	B	20-50	31.6	52.2	16.3	100.0
	C	50 plus	26.6	63.4	10.0	100.0
3. Davidson clay loam	A	0-9	31.9	29.4	34.3	95.6
	B¹	9-36	16.0	22.3	60.4	98.7
	B²	36-60	18.5	30.4	50.3	99.2
	C	60 plus	35.4	34.5	29.6	99.5
4. Carrington silt loam	A	0-12	49.1	36.6	14.3	100.0
	B	12-30	41.0	39.6	19.4	100.0
	C	30-48	44.7	30.1	24.6	99.4

the surface. Thus it is shown only partly in the clay triangle. The Carrington silt loam, another Midwestern soil, lies in the middle triangle, with sand and silt both much more abundant than clay.

Structure is the arrangement of soil particles. Soil structure refers to the manner in which the individual soil particles are arranged. Structure has much in common with texture, although structure is much more complex. As a property of soil, structure in some instances may be even more important than texture. Physical, chemical, and biological forces in nature work together arranging soil particles into a great variety of structural patterns.

Individual soil particles may be grouped together in a great number of different structural forms. Texture, chemical nature, aëration, organic matter, soil water, and other factors complicate the grouping process. Groups of soil particles sometimes are called *floccules,* from the word "floc," meaning

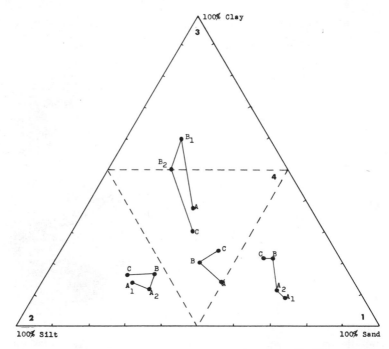

Fig. 1–3. This diagram shows texture graphs for four soils. Note that the three corner triangles are each identified with a particular texture category: 1, sand; 2, silt; 3, clay. The intermediate position of the center triangle (4) associates it equally with each of the three categories. Thus, a graph falling within triangle 1 (Norfolk sandy loam) shows a high sand component. In triangle 2 the graph for Clinton silt loam shows a high silt factor. Graph 3 falls largely within triangle 3 and shows that Davidson clay loam is high in clay. Graph 4 (Carrington silt loam) falls within the center triangle. The sand and silt components of this soil are both relatively high and its clay component is about half that of each of the other two.

Table 1 shows the texture analysis for each of these soil samples. Note that the percentages in the table provide the basis for drawing the respective graphs. Study each graph, referring to the table for percentage figures. Note the texture changes from depth to depth (from A to B to C horizons).

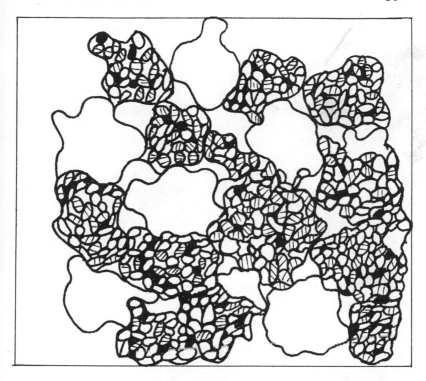

Fig. 1–4. Soil structure is the arrangement of soil particles. This diagram illustrates a pellet, or shot structure. It shows that a well-flocculated arrangement is a very important property of soil structure as it provides ideal conditions for root growth and aeration. A well-flocculated soil is loose and mellow, and easy to till.

the aggregation of fine suspended particles. When once formed, floccules behave somewhat as individual particles, arranging themselves into larger groups. Thus flocculation results in building up a structure with improved aëration, and better opportunity for root circulation and growth. Some structure types have descriptive names, such as "granular," "crumb," "nut," "platy," "massive," and "columnar." Sands are structureless, since each grain behaves independently.

Good structure is a valuable asset to any soil. Some soils have structures that make them difficult to manage and render them practically worthless agriculturally. All soils are vulnerable to abuse; but because of structural differences, some soils require much more care than others. Preventive measures often check structural breakdowns, and careful management can restore deteriorated structures to normal.

The quantity and arrangement of water vary greatly. In supply and in mobility, water is the most variable property of soil. When soil water is abundant, because of either excessive rain or inadequate drainage, all the pore space is filled and the soil is waterlogged. If underground drainage is possible this *gravitational* water moves downward by its own weight into the ground-water zone. As drainage takes place aëration improves, since soil air again fills the pore space vacated by the gravitational water.

After the gravitational water drains away, the soil remains moist because a thick water film, held by surface tension, coats soil particles. This is *capillary* water. When capillary water is abundant it moves slowly downward; when it is diminished by plant use or by evaporation, it may move horizontally or even upward by surface tension. In fine-textured soil, where the space for capillary action (capillarity) is narrow, the upward movement of capillary water may be from depths of several feet. In coarse-textured open soils, the upward movement is more restricted.

When soils dry out, losing their gravitational and capillary water, a limited amount of *hygroscopic* water is held as a very thin film that coats soil particles and colloids. Hygroscopic water does not move from one part of the soil to another, and it is very resistant to evaporation. Thus it is always present in varying amounts, even under near-desert conditions, and provides a valuable reserve.

Soil water plays a variety of roles. The functions of soil water are varied. Soil water is vital to plant life, since all *nutrients* that plants take from the soil must first be absorbed by soil colloids in order for the tiny plant roots to feed. Except in extremely dry soils, these colloids absorb great quantities of capillary water and a limited coating of hygroscopic water, thus making plant feeding possible. Water aids in the decomposition of organic and mineral matter and in bringing about chemical changes within the soil. Water also functions in the movement of materials within the soil, as illustrated by the downward movement of water that transfers dissolved and fine suspended materials from the surface soil to lower levels. This process is called *leaching*. (See p. 18.)

Soil water is a very significant factor in planting, tilling, and harvesting cultivated crops. It often determines the time and the depth at which seeds should be planted for proper germination. Water may be so abundant in the soil as to restrict machine cultivation, thus making the control of weeds difficult. On the other hand, scarcity of water may make the soil hard, cloddy, and very difficult to plow. Too much soil water at harvest time often delays or completely prevents the use of harvesting machinery.

Color is a helpful key to soil conditions. Color is one of the most obvious properties of soils. Soil color is a distinctly conspicuous landscape feature in some areas, whereas in other areas it may be virtually unnoticed. To the person concerned with soils, color is most significant as an indicator of physical and chemical conditions.

Soils range in color from white to black, but the most common colors are the different shades of red, yellow, and brown. These colors indicate the different degrees of hydration and the concentration of iron and aluminum oxides which coat or stain the soil grains. Good aëration is believed by many scientists to be a factor in causing iron oxidation to give a strong red color to many well-drained upland soils in warm, humid climates, such as those of the southeastern United States. These red soils contrast strongly with neighboring yellow soils that have developed under conditions of poor drainage. Some humid-climate soils are whitish or gray as a result of paucity of iron oxide. Light colors in arid climates may indicate a concentration of salts. Dark-brown and black colors usually—but not always—denote a high organic-matter content.

It is commonly assumed, and with good reason, that dark-colored soils are more productive than light-colored soils. This is usually, but not always, the case. Dark color suggests high productivity but does not assure it. Some dark organic soils are strongly acid and unproductive unless heavily limed; whereas other soils are dark because of parent-rock influence and may be poor soils for agriculture.

CHEMICAL PROPERTIES OF SOILS

The study of the vast array of elements and compounds present in the mineral and organic components of soil, and their interaction with each other, is the specialty of the soil chemist. Much of the work done in the soils laboratory involves soil analysis; that is, the testing of soil samples to determine the presence and the amounts of certain elements and the acidity or alkalinity of the sample. Through such chemical analysis and through study in the experimental soil plot, the scientist learns of the needs of soils and develops chemical treatments to meet those needs. Fertilizers are developed and improved. A variety of powders, liquids, and gases are developed for controlling pests. Through the development of "herbicides," chemists are opening up a new system of farming on land that is not suited to ordinary tillage.

A few soil elements are critical. Soils contain most, if not all, known elements in varying amounts and many forms. Oxygen, silicon, aluminum, and iron are the most abundant. Rarely, if ever, does a soil show a deficiency of any of these four elements; and since plants can always secure the limited amounts of them that they require, they are of little concern to us here. However, many soils are deficient in several other elements that are critical to plant growth. These sometimes are referred to as "fertilizing elements," since they are widely used in artificial fertilizers. Nitrogen, phosphorus, and potassium are the three most common. They are constituents of most commercial fertilizers, with their proportions indicated by such formulas as 5-10-5 and 6-4-4.

A few elements essential in small amounts to many plants are contained in very small quantities in most soils. These have been referred to as *trace elements,* because the amounts present in the soil can neither be estimated nor determined very accurately. Their presence can be detected qualitatively, however, and the elements are reported as being available, though in unknown amounts. More recently the term "trace elements" has been partially replaced by the term *micronutrient,* because chemists have concluded that plants need certain elements in specific amounts rather than in "traces." Zinc, boron, copper, and magnesium are "micronutrients," since plants need them in specific, though small, amounts. In fact, some soil biologists insist that the term "trace element" should be reserved for those elements which plants do not require as nutrients. For example, silver and gold are known to be present in very small amounts in some soils. At this time, however, it has not been determined that these two elements are necessary for plant growth; therefore, in a true sense, silver and gold should be called "trace elements."

Soil conditions range from acidity to alkalinity. *Acidity* and *alkalinity* are directly opposite conditions of soil. *Neutral* soils are neither acid nor alkaline. Soil water becomes acid by absorbing carbon dioxide from the air and by absorbing acid products formed by the decomposition of mineral and organic matter. The carbonic acid that is formed from the reaction of carbon dioxide from the air with soil water is a weak acid solution.

$$\underset{\substack{\text{From} \\ \text{the} \\ \text{air} \\ CO_2}}{} + \underset{\substack{\text{Soil} \\ \text{water} \\ H_2O}}{} \longrightarrow H_2CO_3$$

As such, ground water dissolves on contact with the more

Fig. 1–5. A portable kit containing chemicals and color charts is sufficient in determining the acidity of soils. This soil scientist makes numerous acidity tests in routine field work.

COURTESY OF SOIL CONSERVATION SERVICE, U.S.D.A.

soluble bases such as sodium hydroxide. Sodium hydroxide (2NaOH), which is alkaline, reacts readily with the weak carbonic acid (H_2CO_3) of the ground water to form sodium carbonate (Na_2CO_3) and water.

$$2NaOH \quad + \quad H_2CO_3 \longrightarrow Na_2CO_3 \quad + \quad 2H_2O$$

Lime (calcium hydroxide) also exists as a base and reacts with the weak carbonic acid to form calcium carbonate and water.

$$Ca(OH)_2 \quad + \quad H_2CO_3 \longrightarrow CaCO_3 \quad + \quad 2H_2O$$

In the course of these reactions which take place in the soil, the ground water, as an acid solution, removes the bases from the soil. This is the leaching process. Where bases are limited and ground water is abundant and acid, soils of varying degrees of acidity result. Where lime and other readily soluble bases are abundant, or where ground water is limited, soils develop a basic or a neutral reaction. Thus it follows that, in a broad sense, soils in humid climates tend toward acidity, whereas soils in dry climates tend toward alkalinity.

Liming is costly but effective. Most plants, particularly most cultivated crops, will not tolerate a high degree of either acidity or alkalinity. Since most agriculture is carried on in relatively humid climates, acidity is a troublesome and costly problem with many soils. Vast amounts of lime are used to neutralize soil acidity. Some soils of rainy regions, however, are derived from highly *calcareous* (composed of calcium carbonate) parent materials, particularly limestones and calcareous glacial till. These soils remain nearly neutral, or even alkaline, in spite of heavy rainfall. Many of them are among our most productive soils.

The test for soil acidity or alkalinity is of paramount importance in order to determine whether and how much liming is necessary to cultivate a particular plant or plants. Chemically, a soil is acid if a water solution contains more acid ions (hydrogen) than basic ions (hydroxyl), and it is alkaline if the water solution contains more hydroxyl ions than hydrogen ions. If a solution contains the same number of hydrogen and hydroxyl ions, it is neutral.

Not all of the water molecules break up into their component parts at any one time. (The breaking down of molecules into ions—as for example, in the decomposition of the molecule of HOH into hydrogen, H^+, and hydroxyl, OH—is known as *ionization*). As a matter of convenience, the concentration of hydrogen ions is usually expressed symbolically. Samuel P.L. Sorenson, a Danish biochemist, developed a system for expressing the acidity or alkalinity of a water solution. He developed a *pH* scale with numbers ranging from 0 to 14 to indicate relative concentrations. For example, at pH 7, the midpoint, there are the same number of hydrogen ions and hydroxyl ions, and the solution is neutral. Any pH values below 7 indicate the presence of more hydrogen ions, or an acid condition; values above 7 denote the presence of more hydroxyl ions, or an alkaline condition.

We must remember that the pH value is not an indication of all the hydrogen and hydroxyl ions, for all the acid molecules have not ionized. Numerous acid molecules may have the ability to dissociate themselves into ions but have not done so.

If a soil must be limed in order to produce a state of neutrality (pH 7), enough lime must be added to react with the free hydrogen ions and also with the acid molecules, which may ionize at a later date. The liming then causes a state of *equilibrium* to exist, which means, in effect, that there are the same number of hydrogen ions and hydroxyl ions. The lime produces a state of *neutralization* as shown in the following equation:

$$Ca(OH)_2 \quad + \quad H_2CO_3 \quad \longrightarrow \quad CaCO_3 \quad + \quad 2H_2O$$

The alkaline base $Ca(OH)_2$ reacts with carbonic acid (H_2CO_3) to form a neutral salt $(CaCO_3)$ and water. Since a considerable amount of lime must be added to a soil area in order to produce immediate and lasting results, the process is expensive, though very effective.

Soils of different textures may not have the same pH values. The active hydrogen ions are in the water solution and naturally will react first when lime is added. The hydrogen molecules that have not yet ionized are held to the surfaces of the solid particles of clay and organic matter. Since clays and organic matter have large surface areas, the potential acidity would be greater among such fine-textured soils. Sandy soils with a small content of clay and organics would have a lower total acidity than the clayey soils. A good application of lime to these soils may be effective for several years.

THE SOIL PROFILE: THREE LAYERS

Soils exposed to relatively constant environmental conditions slowly acquire many characteristics. Several acquired features distinguish fully developed soils from unaltered, weathered earth materials. One such characteristic is a layered arrangement of materials, called the *soil profile*. The individual layers of the profile are called *horizons*. Most fully developed soils have three horizons, designated, from the surface downward, as the *A horizon,* the *B horizon,* and the *C horizon.* Soils that have not reached full maturity may not have all three horizons represented. The B horizon is the last to develop, and is missing from many soils. In many instances soils have lost one or more horizons through erosion.

A vertical section, such as the fresh wall of an open ditch or a road cut, provides an excellent opportunity to observe and examine the profile features of the soil. Soil surveyors take their observations largely from auger holes, examining the material inch by inch as they drill.

Profile features provide distinction between individual soils. The fine distinctions between individual soils lie largely in their profile features. Soil profiles differ widely in many ways. A generalized drawing (Figure 1–7) shows a wide range of features that might develop in soil profiles. No one soil normally has all these features. Yet if a large number of soils are examined, all these profile features are seen in their relationships to each other. Thus detailed profile features, when carefully interpreted, give individuality to each specific soil.

The surface layer of the soil, or the A horizon, is usually referred to as the *topsoil*; the next layer, the B horizon, as the *subsoil*. The C horizon consists largely of weathered, mineral parent material. Sometimes the underlying bedrock, sand, or other earth material is designated as the D horizon. Since the materials of both the A and B horizons are altered considerably by the soil-forming processes, they are distinctly dif-

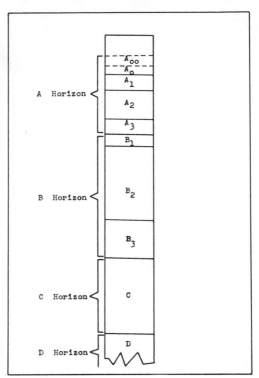

Fig. 1–6. Soil profile showing all the principal horizons. No individual soil has all the features shown here, but all soils have some of them. Most soils have three horizons—A, B, and C—but rarely are all the horizon subdivisions clearly marked. Many soils have lost part or all of the upper horizons through erosion. Very young soils show scarcely any profile features.

The A horizon, or surface soil, is shown here with five subdivisions: A_0 and A_{00} consist largely of surface organic matter; A_1 is a mixture of organic and mineral matter; A_2 contains less organic matter and has a maximum of leaching (eluviation); A_3 is a transition to the B horizon but is more like A than B. The three subdivisions of the B horizon, or subsoil, are: B_1, a transition from A but more like B than A; B_2, the true subsoil with deepest color, maximum accumulation of clay materials (illuviation), and blocky structure development; B_3, transitional to the C horizon, but more like B than C. The C horizon is the weathered parent material: it sometimes has distinct layers. Sometimes the D horizon is recognized; it may consist of hard rock, sand, or gravel lying entirely underneath the parent material.

ferent from the parent material. Together they comprise the true soil, the *solum*.

However, many grassland soils develop only two horizons, the A and the C. These are sometimes called A-C soils. Again, in humid climates erosion may have removed all or part of the surface soil, the A horizon, from many cultivated tracts and denuded forest areas.

The A horizon is the chief root zone. The A horizon of a maturely developed soil can be distinguished from the other horizons by its color, texture, and structure. It is the root zone for most shallow-rooted plants, particularly grasses. Since many roots remain near the surface as plants die, and since leaves, stems, and branches collect on the surface, the A horizon normally is rather heavily charged with organic matter.

Color varies greatly with depth, and from one soil to another. Color usually is influenced by the amount and kind of mineral and organic materials. The mineral part of the A horizon often is darkened by organic stains; but in many soils it is lighter in color than the B horizon. The texture of the A horizon tends to be somewhat coarser than that of the B horizon, although in some soils there is little difference.

The A horizon constantly loses materials by the downward and outward movement of rain water, which carries away dissolved organic and mineral materials as well as fine materials in suspension. Thus the A horizon is *eluviated,* or impoverished, by this loss of materials. As a result, it is made poorer in organic and mineral substances and coarser in texture. In addition to these losses, the A horizon suffers practically all the soil erosion that takes place. This situation seems almost a paradox, since the collection of plant roots, leaves, stems, and branches constantly adds organic material to the A horizon at the same time movement of water takes these and other materials away from it.

The B horizon is the soil storehouse. The B horizon is said to be *illuviated,* since much of the fine material brought down from the A horizon is deposited there. This process of

Fig. 1–7. No two soil profiles are alike. This is a convenient arrangement for display and for classroom use.

enrichment progresses very slowly. It is the last stage in developing a mature soil. Some soils never reach this stage and, therefore, they never develop a B horizon. In the process of forming this horizon, some fine-textured materials are brought in suspension by downward moving gravitational water and are deposited much as material is caught in a filter. Other materials are dissolved in the A horizon, carried downward in solution, and reprecipitated in the B horizon. This horizon, then, might be considered a storehouse for the soil. Indeed, the real strength of a soil for growing plants can be determined largely from the B horizon or subsoil.

The color of the B horizon usually differs from that of the A horizon. In fact, the color classification of a soil is usually determined by the B horizon. Since most of the organic matter has been converted to humus and assimilated in the B horizon, it is frequently darker than the layer above it. Also, oxidation and other chemical processes influence color, as evidenced by the bright red subsoils of many warm-climate upland soils. The B horizon usually contains more clay and other fine material than the A horizon, giving it a finer texture. The B horizon also shows distinct structural development.

As stated earlier, the A and B horizons together comprise the true soil, the solum, since both horizons undergo considerable change as soil forms and become quite different from the original parent material.

The C horizon is parent material. The C horizon consists of parent material that has been changed very little by soil-forming processes. This layer varies greatly in thickness, depending upon the depth of weathering of bedrock or transported material. In smooth uplands of old landforms, the weathering extends tens of feet. On the other hand, in areas of steep slope and in most glaciated areas, the weathering is limited to a few feet. As a soil forms, the C horizon simply supplies the parent material for the A and B horizons. Only the uppermost part of the C horizon is involved in the soil-

Fig. 1–8. A good exposure brings out the layered arrangement of soils. This photograph shows three distinct layers. Also, it shows that the layers do not always meet along smooth surfaces. Tongues of one horizon often extend deep into the next horizon. (Kendellville loam; Shelby County, Indiana)

COURTESY OF SOIL CONSERVATION SERVICE, U.S.D.A.

forming processes as the solum above it becomes deeper with time.

Environmental changes are recorded in the soil profile. The soil profile provides perhaps the finest record of the soil's natural environment, past and present. It shows the conversion of weathered rock material into the dynamic natural substance we know as the true soil, and so preserves the story of the interaction of all natural environmental factors—climate, vegetation, topography, drainage, and time. Slow changes in climate take place through pedologic time. (See Chapter 2,

p. 39). These changes reveal themselves in features of the soil profile. This is true because climatic changes usually result in changes in vegetation, in chemical reactions within the soil, in drainage conditions, and in kind and extent of erosion. For example, a study of a soil profile in New Jersey showed that the eastern United States has experienced three distinctly different kinds of climate since the last glacial period, which ended some eleven thousand years ago.

DEVIATIONS WITHIN THE PROFILE: PAN LAYERS, CALCIUM LAYER, SHRINKAGE AND SWELLING

The pan layer is not a profile feature in the usual sense. It is a dense, compact, or cemented layer resulting from a number of causes common to soils that otherwise have normal profiles. In humid regions pan layers sometimes result from excessive ground water; in dry regions the precipitation of carbonates sometimes causes pan layers.

Hardpan soils may result from drainage problems. *Hardpan* is a type of pan layer involving cementation, usually in the lower part of the B horizon. The cementing agents are derived from different mineral compounds and from organic matter. Iron oxide is a very common mineral compound involved in hardpan formation in soils of humid climates. Layers cemented with iron oxide sometimes are referred to as iron crusts. In some soils these iron crusts are broken up by some natural process, and ironstone fragments become distributed throughout the profile and on the surface. Calcium carbonate and silica compounds also act as cementing agents in some hardpan soils. Organic hardpans are perhaps more widely distributed than mineral hardpans in humid regions. They occur over much of the Atlantic and Gulf Coastal Plain of the southeastern United States. Such organic pan layers are associated with a high level of ground water. They generally are located within two or three feet of the surface, at the level at which water usually stands. Hardpan soils are usually of rather limited agricultural value. Since plant roots cannot readi-

ly penetrate the pan layer, only shallow-rooted plants can adapt
to them. Citrus groves do not grow where soils of Florida are
hardpan. In North Carolina, however, most of the blueberry
crop grows on hardpan soils.

Claypans are often natural; they may be induced.
The *claypan* of some soils is a type of pan layer resulting
from compaction or from excessive concentration of clay in
the B horizon. No cementation is involved in claypan formation. Little change other than compaction and concentration
takes place in the nutrient component or in other properties
of the B horizon clay. Claypans are widely distributed, and
vary greatly in depth, thickness, and permeability. Excessive
use of heavy farm machinery is one of the major causes of
claypan formation. Several methods are used to break up claypans, of which deep plowing is perhaps the most successful.
Claypans, like hardpans, restrict the uses of soils. Good farm
management involves finding special uses that employ the claypan to advantage rather than attempting to destroy the pan
layer entirely (on land in which it is a problem).

**Pan layers check root growth and impede drain-
age.** Hardpans and claypans greatly reduce the value of soils
for agricultural uses. This is largely because they interfere with
plant growth and impede downward drainage of excessive
water. Not only do hardpans restrict crops and orchard trees
in some areas, but they often limit natural forest growth to
stunted, worthless trees. When the pan layer is relatively deep,
it will not interfere with the growth of shallow-rooted crops
if the land is drained by ditches or tile. In fact, the pan layer
may sometimes be an asset, as in the case of rice growing
on some soils in which the pan layer serves to hold irrigation
water near the surface in reach of the rice roots.

**Calcium layers develop in the lower solum in dry
climates.** One of the distinguishing characteristics of grassland soils in semiarid and arid regions is the "lime layer"
that forms in the lower part of the solum. The presence of
such a lime layer assures an abundant supply of calcium and

thus eliminates acidity problems. Thus the presence of a lime layer is a salient feature of some of the world's foremost wheat and pasture lands. The process responsible for the formation of this lime layer is called *Calcification*—a process that will be discussed in detail in Chapter IV. In general, Calcification takes place at approximately the depth of rainwater *percolation*. On the rainier margins of the regions in which the lime layer is formed, it lies from three to five feet beneath the surface; on the drier margins, however, it lies only a few inches underground.

Artificial drainage lowers the ground level of some soils. The water content of many wet-land soils is extremely high. This is especially true of organic soils, classed as peat and muck. When such soils are drained by tile or open ditch systems, the drained portion that lies above the tile or ditch level shrinks as a natural result of water withdrawal. In some soils, depending upon the amount of organic matter present, shrinkage may lower the ground surface by as much as a foot. This situation can produce complicated problems in areas that lie approximately at sea level.

The swelling of the subsoil helps stabilize soil water. Some soils in low-lying, flat areas of humid regions have a B horizon of heavy clay that possesses the unusual quality of swelling upon becoming wet. Such a subsoil is virtually impervious to downward percolation water. The watertight subsoil holds irrigation water in the root zone near the surface, preventing its downward escape. This may prove to be a good soil quality in irrigated areas, such as the rice lands in Louisiana.

Word List for Study

mantle	muck
floodplain	transported materials
pedologist	oxidation

alluvium   decomposition
regolith   texture
organic materials   colloids
residual materials   floccules
parent material   gravitational water
glacial till   capillary water
eolian   nutrient
loëss   hygroscopic water
peat   trace elements
humus   neutral
alkalinity   micronutrient
calcareous   acidity
pH factor   leaching
profile   ionization
topsoil   equilibrium
solum   neutralization
illuviation   soil horizons
subsoil   eluviation
claypan   hardpan
Calcification   percolation
raw humus

Questions for Consideration

1. What is the significance of the top six inches of soil in the human habitat?
2. Deserts may have some areas with deep mantles of shifting sand and other areas of bare rock. Why?
3. How do you account for forest growth on mountain sides?
4. Describe a river floodplain. How may the depth of soil vary from year to year?
5. What steps can be taken to reduce the problem of erosion on steep slopes?
6. What makes soil? Explain the importance of air in the soil.

7. Where does the organic material in the soil come from?
8. What are the sources of parent materials?
9. What is meant by the term "waterlogged soils"? What kinds of plants can grow under these circumstances? Explain.
10. List the physical properties of soils. Explain each one briefly.
11. Why are the loams considered to be the best soils?
12. Explain why careful management of the soil is so important to us today.
13. Distinguish between gravity and capillarity in the transfer of soil water.
14. Do soils have pores? Would good productive soils have a good porosity? Why or why not?
15. What is the importance of soil water in planting, tilling, and harvesting crops?
16. Explain the range in soil colors. What do the different colors represent?
17. List some critical elements necessary for plant growth.
18. What is meant by soil acidity? Alkalinity? Explain how tests can be made.
19. What is liming? Why is it so expensive?
20. Discuss the neutralization process. What is the significance of a pH test?
21. What factor in addition to the pH number must be considered if liming is to have lasting results?
22. What is a soil profile?
23. Explain why soil horizons differ in various sections of the country.
24. What does a soil profile tell about natural history?
25. Why is it difficult to develop an orchard in areas with a hardpan?
26. Explain why some soils shrink or swell.
27. What is meant by "ground fires"?

CHAPTER 2

The Factors that Affect Soil Formation

MATERIALS CHANGE AS SOILS FORM

The materials of which soils are composed exist in other natural forms, but soils differ from other forms and from one another. Differences among soils reflect the fact that all mature "true" soils develop as a result of particular combinations of causal factors and processes. This chapter concentrates on the causal, or genetic, factors, while Chapter 3 deals with the soil-forming processes.

The variety of materials found in soils and the variety of soil-forming factors and processes are so great that the potential number of individual soils, each with its own identity, is virtually limitless. For scientific purposes, therefore, the general term "soil" is not specific enough. A single farm may have as many as twelve different soils, a single county a hundred or more, and a still larger area, such as a state or region, several hundred.

Individual factors account for specific changes. Soil formation is slow and so intricate that the isolation of cause-and-effect relationship is difficult and sometimes impossible. However, a few broad, general conclusions, based on careful scientific observation, can be of help in interpreting soils in their natural setting. For example:

† In warm, humid climates leaching is relatively active and soils tend to develop an acid reaction.

† In most soils in areas of poor drainage, organic matter is not well assimilated.

† In upland regions with grass vegetation, high organic content and dark color characterize many soils.

† Fresh deposits of river alluvium or dune sand lack
profile features because of insufficient time for soil development.

These examples, of course, present only limited, isolated aspects of individual soils.

Soils vary greatly in extent of development. In
some soils it is apparent that nature has done very little to
alter the body of weathered earth (parent material) from
which the soils have been formed. Such soils have barely
begun their evolutionary development and are just entering
the first stage of their course toward maturity. Other soils have
changed so much since the parent material first accumulated
that identification of the original parent rock is very difficult

Fig. 2–1. This very shallow Lithosol consists of a thin layer of weathered
rock material resting immediately on the bedrock. It shows very little
profile development, and it supports only sparse vegetation. (Sloan series;
Gage County, Nebraska)

COURTESY OF SOIL CONSERVATION SERVICE, U.S.D.A.

or impossible. Obviously, several factors are responsible for these changes and some factors are more effective than others in bringing them about in any given soil.

FIVE GENETIC FACTORS OF SOIL FORMATION

Pedologists generally agree that soils are evolutionary in nature, passing through stages of development that are comparable to the life cycles of natural vegetation and animal life. There is less agreement, however, concerning certain details of soil formation. One of the points of disagreement has to do with the degree to which each individual genetic factor is involved in soil formation. The more we learn about soils, the more complex they appear to be, and the less certain we can be about how they were formed.

The science of pedology generally recognizes five genetic factors that contribute to the formation of mature virgin soils: (1) parent material, (2) plant and animal life, (3) climate, (4) relief (particularly as related to drainage), and (5) time. Regardless of parent material, virgin soils become more and more alike as climate, vegetation, slope, and drainage remain more or less constant over a long period of time. Like plant associations, soils within a given area reach full maturity when they have come into a state of equilibrium or balance with these environmental factors.

Man as a modifier of his environment (including soils) might also be considered a genetic factor of soil formation. But man is not a genetic factor of virgin soils, since they acquire all their characteristics before man enters upon the scene to change them. All agricultural soils, no matter how much they have been modified by man, were virgin soils before they were cleared for cultivation, but no virgin soil remains in its natural state after cultivation begins. Some are greatly damaged, others are improved, but all are changed in some degree, by man's use of them. In many areas of long continuous use, it is often difficult to find a virgin soil with which to compare the soils under cultivation. It may seem a paradox to emphasize the natural state of virgin soils in this book, since we are dealing in a large measure with soils in agricultural use regardless of changes that man has brought to them. But it is necessary to understand the natural factors and processes of soil formation in order to interpret the different ways in which various soils respond to man's treatment.

ROCKS PROVIDE PARENT MATERIAL: WEATHERING ALTERS IT

Soil parent materials differ as widely as do the rocks of the earth's crust. Many forces cause rocks to disintegrate and to mantle the solid earth with debris of varying depths. Geologists call these the *gradational forces*. (*Degradation* degrades, or wears away, while *aggradation* builds up through accumulation.) *Erosion* is the common term for wearing away and removal. Running water, wind, glacial ice, and gravity are the chief agents of erosion and deposition. Even while the accumulation of deposits is taking place, disintegrated mantle rock undergoes further changes through the processes of weathering. Among them are decomposition and other chemical changes. Thus the parent material of a given soil may be quite different from rock that has disintegrated but remained the same in all other respects.

Parent material provides bulk and essential elements. The parent material provides not only the bulk of a soil but also determines, largely or entirely, that soil's composition with respect to elements required by plants. In a particular soil some essential elements may be in short supply, while others are abundant. Some parent rocks, such as certain limestones, indicate unusually good agricultural soils, whereas shale and sandstones suggest soils of poorer quality. The parent material also determines, within broad limits, such physical properties of a soil as its texture, its structure, and its water capacity.

Much parent material is transported. Materials transported and deposited by streams, glaciers, and wind cover much of the earth's surface to varying depths. Often these deposited materials are derived from a variety of rocks, and are thoroughly mixed.This is also true of materials that blanket the floors of extinct lakes and of lakeshore plains. The parent materials of these transported deposits give rise to a variety of soils. On river floodplains, where each flood deposits a new layer of mud, soils derived from recent alluvium have limited opportunity to develop toward maturity. Despite their immaturity, however, many alluvial soils can be highly productive if they are properly drained.

PLANT AND ANIMAL LIFE FURNISH LITTLE BULK, BUT ARE HIGHLY IMPORTANT

Organic material comprises most of the solid matter of soils that is not parent material. Parts of plants furnish most of the organic material. Vegetation, a highly significant physical soil property, is no less significant as a genetic factor of soil formation. The fact that forest soils, grassland soils, and desert soils differ greatly from each other is common knowledge. The interpretation of these differences, however, requires proper evaluation of vegetation as a genetic factor.

Organic matter plays many roles. Organic matter has many functions in soils. Some of the essential nutrients for plant growth come directly from organic matter. Where it is insufficient, artificial fertilizers must be used to supplement it. Organic material contributes chemical constituents to soil water largely in the form of acid products. These chemicals perform several functions, but perhaps the most important is to break down some of the minerals of the parent rock, making them available to plants.

Organic matter is an important factor in the water capacity and rate of absorption of some soils. Often "droughty" soils are greatly improved by generous additions of organic matter. Tight "clammy" soils with poor air circulation show marked improvement when organic matter is added. This is particularly important in the early stages of plant growth, at the time of

Fig. 2–2. Prairie soil, Sharpsburg silty clay loam series. Note the dark color and high organic content of the A horizon of this Midwestern grassland soil.

COURTESY OF SOIL CONSERVATION SERVICE, U.S.D.A.

seed germination and early root development. Soils with abundant organic matter show little tendency to form "crusts" when they dry out after rains; therefore, the young tender plants have little difficulty in breaking through the surface. In cultivated soils organic matter is an important factor of tillage. Not only does it contribute to the ease of cultivation, but it also reduces the delay of tillage after rains.

Organic matter reduces soil erosion.This is of particular significance where abundant undecomposed material remains on the surface and where plant roots form a tangled mass in the A horizon. Decomposed organic matter, or humus, also tends to bind together soil mineral particles, induce water absorption, check the rate of water runoff, and reduce the removal of surface soil particles by wind. Proper management and control of organic matter is one of the best means of preventing both water and wind erosion.

Vegetation leaves its record in the soil. In most virgin soils vegetation stands out clearly in the profile. Peat, muck, and other soils that are high in organic matter provide obvious examples. Grassland soils reveal their organic matter by the presence of a dark color and by the fibrous nature of the A horizon. In forest lands the distinction between the *deciduous* and the *coniferous* growths of the original cover can usually be clearly seen even after the timber is removed. Soils under cultivation show the imprints left by natural vegetation for years after the virgin growths have been destroyed and the agricultural crops have been planted. Sharp contrasts may be seen today between the black prairie soils, some of which have been in cultivation for more than a century, and the gray-brown soils of the forested uplands of Illinois.

Many soils are known to have developed under a type of vegetation that was different from that of modern times. Slow changes in climatic conditions account for many changes in the natural vegetation. Changes in drainage conditions also account for many changes in vegetation. Soils reflect these changes in their profiles. A good illustration is found in the plateau uplands of Spain where the vegetation is short grass. The soils are believed to have developed under a dry forest that existed many centuries ago when rainfall was somewhat more plentiful than it is now. Studies of soils in various parts of the world reveal that throughout pedologic time changes in vegetation have accompanied changes in climate and in drainage.

CLIMATE SETS THE STAGE FOR SOIL DEVELOPMENT

Rocks and vegetation provide the materials of which soils are made but climate governs soil development. Climate and soils tend toward a zonal arrangement. Just as climatic regions show a strong influence of both temperature and *precipitation,* so also does the broad pattern of soil arrangement. In humid lands, such as those in the eastern part of North America, temperature is the dominant climatic factor in soil differences. In the humid eastern region, soils are arranged in north-south zones according to temperature. By going from the rainy eastern part to the dry interior of the United States, we can see soil is zoned according to rainfall. This broad relationship between soils and climate can be seen in the zonal patterns they assume over the entire earth. (See the maps in Chapter 5.)

Climate exerts both direct and indirect influences. The climatic factors in soil formation are complexly interrelated. Both directly and indirectly, climate influences soil formation in many ways. It is not generally possible to measure accurately the part played in soil genesis by each specific climatic element. Climate itself is highly complex—the soil climate may differ greatly from the atmospheric climate in a given area.

Temperature and moisture contribute more to soil formation than do other climatic factors; their different imprints on the soil profile often may be identified and measured. However, the two tend to work together in affecting soils. In general, as temperature and precipitation increase, the extent and speed of their influence on soil formation increases. In the hot, rainy tropics the climatic factor in soil formation is most pronounced and its results are most rapid.

As we have already pointed out (see p. 36), the indirect results of climate on soil formation are most clearly seen in the relationship of soils to vegetation. Drainage also is related to climate in many ways. The amount, the seasonal distribution, and the intensity of precipitation are factors in soil drainage, since they affect surface runoff and water penetration into the soil. Temperature affects drainage, particularly when the ground freezes and rain or snow cannot penetrate the soil. Wind also is a climatic factor affecting drainage in that high wind velocities increase the amount and the rate of evaporation from moist soils. The relation of drainage to climate, however, is usually less pronounced than is the relation to elevation and slope. The drainage factor in the genesis of soils is evident under a great variety of circumstances. Where poor drainage is corrected artificially, many soils become highly productive, yet soil qualities associated with poor drainage, such as yellow or gray color and high organic content, remain long after artificial drainage is provided.

Climate regulates chemical and biological reaction. Many chemicals are more active in relatively high temperatures than in low temperatures. In the tropics, with high temperatures and heavy rainfall throughout the year, leaching of soluble materials from the upper part of the soil is rapid. Organic matter decays swiftly, releasing materials that increase the dissolving properties of soil water. Biological reactions are also swift. Abundant animal life including ants, worms, and burrowing animals devour the dead trunks, branches, stems, and leaves of plants, thus reducing the amount of organic matter returned to the soil.

In cold climates, low temperatures retard chemical reaction in soils. But leaching, just as in the tropics, is pronounced; so much so that it gives soils a distinct acid reaction. Decomposition of organic matter is slow. Fungi activity increases, but bacteria are less active than in soils of warmer climates. It must be borne in mind that the soil is frozen to considerable depths for much of the year and that practically no change—chemical, biological, or physical—can take place under frozen conditions.

Another principle to keep in mind is that most chemicals react slowly or not at all unless they are moistened. Lime and other soluble materials are not removed from parent material by natural processes unless they are dissolved by ground water.

Fig. 2–3. The Red Desert soil of New Mexico is found widely in our arid West. Note the very light color due to scarcity of organic material. The scarcity of vegetation and the nature of the growth suggest the desert climate.

Similarly, the decomposition of mineral and organic material is retarded in the absence of water. In very dry climates, therefore, decay is slow, leaching is at a minimum, and some highly soluble minerals, particularly calcium carbonate, salts, and alkalies, become concentrated in the soil. Because of these conditions many dry-region soils are high in most minerals essential to crops and, unless harmful salts are abundant, they are very productive when they are irrigated.

RELIEF EMBRACES BOTH SLOPE AND DRAINAGE

The significance of relief as a genetic factor of soils is more noticeable locally than over large areas. Most frequently, the soils are shallow on steep slopes because erosion is accelerated. There are exceptions, however. In some areas of slight relief and gentle slope, bedrock may appear at or very near the surface, with little weathered material on top of the rock. In glaciated areas, many level surfaces were scraped bare by advancing ice and afterwards were never covered with unconsolidated material. On the other hand, in some areas of relatively steep slopes there are thick blankets of weathered material that give rise to deep soils with mature profile features. Many of these areas have remained protected against erosion by dense forest cover and have been handled carefully after clearing. As a result the soils have remained good farming lands. Nevertheless, steep slopes tend to have shallow, stony soils with poorly developed profiles.

Relief affects the degree to which soils advance toward maturity. Rapid erosion removes surface material before ma-

Fig. 2–4. Mountain farm scene in western North Carolina. Only the small field of Burley tobacco remains; the rest of the land is abandoned from crop production. Steep slope and associated erosion has brought this about. This Ramsey soil series is derived from shale and seldom is very good for crops.

COURTESY OF SOIL CONSERVATION SERVICE, U.S.D.A.

ture profile features have time to develop, so that soils in areas of rugged relief usually show youthful features without the full marks of maturity in their profiles. We must not overlook the fact, however, that for normal development of soils, some slight to moderate slope is necessary. Gradual removal of surface material by normal erosion has a rejuvenating effect that prevents soils from becoming old and sterile. Level land often has impaired drainage that results in retarded soil development.

Relief and soil drainage go hand in hand. Steepness of slope accelerates soil drainage as it does erosion. (The effectiveness of close spacing of terraces on steep slopes as a means of slowing down excessive drainage illustrates the point very well.) Steep slopes and accelerated erosion often cause a slow downward movement of the thin soil layers under the force of gravity. This process is known as *soil creep,* and produces *colluvial soils* (mixed deposits at the base of slopes).

The catena expresses local drainage differences. Relief and drainage provide a basis for convenient grouping of local soils. Different individual soils that are derived from identical parent material and have developed under environmental conditions which are similar in all major respects except relief or drainage sometimes are grouped in a sequence based on drainage. Such a grouping is called a *catena*. This is an appropriate term since catena means "chain." Each individual soil of the catena represents a link in the chain of several related soils.

The catena concept provides an excellent approach to studying soils in the field with the aim of seeing the relationship of individual soils to each other and to such environmental factors as relief, drainage, parent material, and vegetation. This approach is an important part of *soils geography*.

TIME SETS THE EXTENT AND FINAL LIMITS OF ALL SOIL DEVELOPMENT

All factors of soil genesis are present in a period of time, but time is not measured in years. In soil formation, as in most other earth processes, nature works slowly. As a parallel of historic time and geologic time, we refer to time that nature has devoted to the formation of soils as *pedologic* time. It might be estimated in years, in centuries, or in milleniums. Since soils tend to develop in cycles, the length of time required for a given soil to complete a cycle, or the length of time that it has already spent in a cycle, is of less significance than the stage in a cycle—youth, maturity, or old age—in which we find it.

Soils vary greatly in rates of aging. All soils do not age at the same rate. In general, aging is more rapid in warm, moist climates than in cold, dry climates. Generally soils age faster on gentle upland slopes than on flat lowlands or on steep uplands. Soils from sandy parent material usually advance toward maturity faster than do soils derived from clay. No set number of centuries can be ascribed to a given soil cycle or to a particular stage in the cycle. Soils on a river floodplain remain young as long as flood waters inundate them periodically. Periodic flooding may continue for thousands of years, but eventually, as streams slowly cut their channels deeper, floods come less frequently. Finally, the soils that formerly were on the floodplain are left on terraces, above the reach of flood waters. They then experience a normal process of development and may advance to maturity within a few centuries.

In the areas of northern United States which were covered by the last continental glacier, soils have had only approximately ten thousand years to develop since the ice disappeared. This may seem like a long time, but in cold climates soils develop very slowly. Many of these soils are not yet fully mature. A recent study was made of a slope near the foot of Mendenhall Glacier, Alaska. From a study of the vegetation, and from other evidence, it was found that the glacier had deposited the material on the slope about a thousand years ago. The study revealed that soil profile features were beginning to develop, although the soil was in the very youthful stage of its cycle, and that full maturity may well lie many thousands of years in the future.

Continual changes take place. Continual changes occur in virgin soils. Most changes come too slowly to be perceived by any one individual, but some striking illustrations are offered by studies of soils that have developed over man-

Fig. 2–5. Poor drainage is characteristic of many of the poorly developed, dark soils of the Atlantic Coastal Plain. Note the broad V-shaped ditch made to drain this Portsmouth series which is found widely in eastern North Carolina. These soils are moderately productive when properly drained, fertilized, and limed.

made structures. A soil has been found and studied that developed on the roof of a buried building in an abandoned Chinese town. The profile study revealed much of the natural history of that area after the town was abandoned. Another example is a youthful soil in Nebraska with some clearly marked profile features that developed on a railroad fill in the short span of 75 years. Occasionally changes are swift. For example, in the early 1800's a large lake, Reel Foot Lake in western Tennessee, was formed by a major earthquake. (Man himself, of course, works swift changes; for example, the many large areas of excellent agricultural soils that have been inundated to form artificial lakes and reservoirs.)

We cannot wait for more soil to form. Regardless of whether nature requires a hundred years or many thousands of years to make a soil, the process is a slow one. We do not have time to wait for nature to repeat her performance, once we have allowed our present soils to waste away. Thus, it behooves all of us to guard diligently this most precious natural heritage in the interest of our generation and of posterity.

MAN'S ROLE IS BRIEF, BUT SOMETIMES OF GREAT CONSEQUENCE

During the many centuries that man has used soils he has modified them in a variety of ways. Although his genetic role is largely one of altering the soils themselves, man also has exerted considerable influence over the natural factors that affect the soils.

Man's role as a soil modifier is seen most clearly in agricultural lands, particularly those used for crops. As more and more land is taken for nonagricultural uses, such as excavation, grading, and filling, an ever-increasing acreage of natural soil is completely destroyed. Artificial reservoirs take millions of acres of good valley soils out of agricultural use. Strip mining is another significant spoiler of soils. In addition to such deliberate removal of land from agricultural use, man is also unintentionally responsible for vast changes in many soils of farms and grazing ranges which are never again used for cultivated crops. Overgrazing, which removes the protective vegetable cover, accounts for wind and water erosion of vast pasture areas. Careless forestry practices bring about the deterioration of woodland soils through erosion and burning. Floods, often man-induced through deforestation and improper agricultural processes, bring vast changes to countless acres. On the credit side of man's activities, proper artificial drainage of wet lands adds vast areas of fertile agricultural land. Fertilization and good management result in the improvement of countless acres.

Man's waste exceeds his improvement of soils. On the whole man has damaged soils more than he has salvaged or improved them. Improper handling, perhaps, deserves most of the blame for man's role as a spoiler of soils. Some of the early cultures sank into poverty and retrogressed largely because their life-giving soils were allowed to deteriorate through improper handling. In America millions of once-productive acres are well on their way to destruction, and man-induced damage continues to gnaw at the heart of most cropland that is left. Land abandonment proceeds faster than the addition of new crop acreage, and the total shrinks yearly. Soil erosion, most of which is within man's power to control, is by far the major agent of soil deterioration. It perhaps is our number one way of wasting this natural resource. Depletion of supplies of plant nutrients, loss of precious organic matter, and injury to soil structure are other damages that can be repaired only slowly and at great cost. Widespread use of artificial fertilizer is, perhaps, the greatest contribution made in the twentieth century to agriculture, although improvements in soil handling and management account for the success of fertilization. In many areas the yields per acre continue to rise. On the whole, mankind is thought by some writers to be better fed than ever before.

A major challenge lies ahead. The damage man has caused to one of the most precious of all natural resources, modified by the improvements he has wrought, constitutes his record as a soil modifier. But, lest we be too critical, we must remember that in using soils in some parts of the world man has fed and clothed himself through more than forty centuries of agriculture. He now faces the challenge of feeding an exploding world population that will probably double in little more than a generation.

The ways that soils will serve the needs of the farmer in the years ahead depend largely upon the advancement of scientific agriculture. Soil science must contribute more and more to the solution of intricate problems. Already scientific agriculture faces the two-fold challenge of conserving our soils and of making them more productive. The American farmer shows strong prospects of meeting this challenge.

Word List for Study

nutrients	alluvium
genesis	degradation
leaching	floodplain
aggradation	absorption
shale	deciduous
organic matter	precipitation
coniferous	colluvium
decomposition	erosion
relief	soils geography
catena	deterioration
terrace	conservation

Questions for Consideration

1. Explain why soils differ from one another.
2. Why is "soil forming" considered to be a slow process?
3. List the five genetic factors that affect soil formation. Explain each one briefly.
4. How does the science of pedology differ from geography?
5. What role does man play in modifying soil formation and soil fertility?
6. Distinguish between the forces of degradation and the forces of aggradation.
7. What do shales and sandstones indicate about the quality of soils?
8. Discuss maturity of soils. Find and give examples of soils which indicate youth, maturity, and old age.
9. What part does drainage play in the formation of soils?
10. Explain the importance of organic matter in soils.
11. What do we mean when we say that soils are arranged in zones?
12. Explain the relationship between the speed of chemical reactions in the soils and the temperature.
13. Why is the productivity of many soils in dry climates high if they are irrigated?
14. In general what is the relationship between the soil profile and slope?
15. How are colluvial soils formed?
16. In a paragraph or so, explain your concept of the catena in classifying soils.
17. Investigate the advantage of terracing a field. Is it always necessary? Why or why not?
18. How can overgrazing affect soils in place?
19. Write a short paper on how man has and is wasting the soil. What may happen to the world's population if he continues to be so careless?

CHAPTER 3

Processes of Soil Formation

PEDOLOGIC PROCESSES AND THE NATURAL ENVIRONMENT

The preceding chapter was concerned with the genetic factors of soil formation. Chapter 3 illustrates the *pedologic* processes in nature that operate to produce soil. It might be said that the genetic factors set the stage, and the pedologic processes act to produce true soil. The pedologic processes are different from all other natural processes. For example, geological weathering produces weathered rock material while only the operation of pedological processes changes weathered rock into true soil.

Most natural processes work slowly; pedologic processes are not exceptions. Geologists estimate the age of the earth to be more than six billion years, and various kinds of changes in the earth have been going on during this geologic time. Some natural changes, such as those resulting from earthquakes and volcanic eruptions, come about rapidly, sometimes suddenly. Most major earth changes, however, such as the carving of a canyon, the uplift of a mountain mass, or the filling of an inland sea, take place very slowly.

By contrast, the pedologic processes, though slow in terms of human life or even cultures, work somewhat less slowly than geological processes in changing lifeless parent material into true soil. As was pointed out in Chapter 2, some soils of the northern United States have advanced far toward maturity since the retreat of the continental ice sheet some ten thousand years ago. Under more favorable climatic conditions, soil formation is even more rapid. The time cycle is never the same for any two soils.

End results are hard to measure. The end results of pedological processes are difficult to measure, for not only is each pedologic process extremely complex, involving many chemical and biological reactions, but usually two or more processes operate simultaneously in a given area. One process may counteract another or two different processes may achieve the same effect.

As we examine each pedologic process in a simple, nontechnical way, it becomes clear that each process is closely related to the natural environment. The relationship between the pedologic processes and the genetic factors of soil formation is very obvious. All five genetic factors contribute in some fashion to the development of each mature soil, but no single soil is influenced by all pedologic processes. Different processes or combinations of processes may perform under varying conditions. For example, grasslands differ from forest lands, humid lands from drylands, and steep lands from level lands. On the other hand, a pedologic process is distinguished by the way it works, not by the kind of soil it produces. For example, a given process operating in a limestone area produces different results from those of the same process operating in a sandstone area (different parent material). Likewise, a given pedologic process may be closely related to other processes. The results obtained from processes operating in combination usually differ markedly from those of a given process working alone. This principle is illustrated over and over when individual soils are classified and studied.

Seven processes account for all soils. Altogether there are seven pedologic processes, each with a technical name of very restricted use. They are: (1) *Calcification,* (2) *Podzolization,* (3) *Latozation,* (4) *Salinization,* (5) *Solonization,* (6) *Solodization,* and (7) *Gleization.* The first three are largely responsible for the mature soils in areas where all genetic factors are in equilibrium with the environment. The last four are involved with immature soils. In some instances they represent transitions from one stage to another in the cycle of soil development.

THREE PROCESSES OPERATE WHERE EQUILIBRIUM ATTAINS; THEY PRODUCE ZONAL SOILS

Three processes, Calcification, Podzolization, and Latozation are largely responsible for the formation of *zonal* soils, that is, the mature soils that have adjusted fully to their environment. If genetic factors are stabilized for a long period of time, the full impact of the environment is reflected in the features of the soil profile. Since zonal soils represent complete adjustment to environment, a brief examination of these three processes helps in understanding the relation of soils to environment.

CALCIFICATION IS THE SIMPLEST OF THESE PROCESSES

Of the three processes, Calcification is the least complex and easiest to understand. Its name is more descriptive than those of the other pedologic processes, since it obviously indicates that calcium carbonate is involved in the reaction. The Calcification process results in the concentration of calcium in the soil. Calcium is readily soluble in acid soil water, and its chemical behavior is not complicated. As calcium is dissolved from one part of the soil and transferred to another part, only simple principles of chemistry are involved, as the following explanation shows.

When calcium is deposited in the upper part of the C horizon or the lower part of the B horizon, the calcium reacts with water to form a lime deposit and to release hydrogen ions.

Water (H_2O) is composed of two atoms of hydrogen and one atom of oxygen. Liquid water is really a mixture of $H+$ and $\overline{OH}$. The $H+$ is the positively charged hydrogen atom and is called the hydrogen ion. The $\overline{OH}$ is called the hydroxyl ion and is negatively charged.

The hydrogen ion and the hydroxyl ion come from the breakdown of ionization of the water molecule:

$$H_2O \; \rightleftarrows \; H^+ \; + \; \overline{OH}$$

All the water molecules do not break down into H^+ and $\overline{OH}$ because the molecules are very stable; in fact only about one molecule in 10 million is ionized at any one time. The more water moving down through the soil, the faster the process.

The equation for the deposited calcium and water is as follows:

$$Ca \; + \; 2H_2O \; \longrightarrow \; Ca(OH)_2 \; + \; 2H^+$$

The hydrogen saturated soil is then neutralized by the lime.

Fig. 3–1. Lime accumulating soils are widely distributed in middle latitude grasslands. This Chernozem (Crete silty clay loam, Republic County, Kansas) is excellent for wheat and for cattle pasture.

COURTESY OF SOIL CONSERVATION SERVICE, U.S.D.A.

Fig. 3–2. This Brown Soil profile is a Clovis fine sandy loam located in Roosevelt County, New Mexico. It is a good grazing soil and is used for crops where irrigation is feasible; however, it is restricted to rather shallow-rooted crops by the thick lime layer 2½ feet beneath the surface.

Soil $\diagdown_H^H$ + Ca(OH)$_2$ $\longrightarrow$ Soil—Ca   +   2H$_2$O

The effect is that the acidity has been neutralized and the soils developed by Calcification have been made neutral or slightly alkaline. After this process the calcium reacts with a weak carbonic acid (formed when water comes in contact with the carbon dioxide in the soil—H$_2$O + CO$_2$ $\longrightarrow$ H$_2$CO$_3$) to form lime or calcium carbonate—CaCO$_3$.

Ca   +   H$_2$CO$_3$ $\longrightarrow$ CaCO$_3$ +   H$_2$↑

The formation of the lime carbonates is of prime importance in the process of Calcification.

Calcification typifies middle-latitude grasslands. Calcification is most typical of grasslands with restricted rainfall. Grasses and grass-like plants are heavy users of calcium bases and other bases. They bring large quantities from the lower soil levels and, on dying, leave them in the upper layers. Capillary water also brings bases upward in the soil where rainfall is light. Thus, these bases tend to accumulate near the surface from depths well below the ordinary root zone. Under conditions of heavier rainfall complete removal of these bases would result from leaching. With light rainfall, however, the dissolved material is carried downward only a few inches and reprecipitated. (Chemical *precipitation* is the separating out of a substance from a solution because of a chemical change or a physical change, as of temperature. It is also the formation of an insoluble compound in a solution.)

The deposition of these bases forms a lime layer, or crust, in the lower part of the solum. Usually it is about at the average depth of percolating rainwater. Its location depends upon (1) the amount of rainfall, and (2) the texture of the soil. Its depth ranges from five feet or more in sandy soils on the rainier margins of grasslands to less than one foot in tight soils of desert margins. Soils with such a lime layer formerly were called *pedocals,* a term formed by adding the "cal" of "calcium" to the root of "pedology."

PODZOLIZATION

In many respects Podzolization is the negation of Calcification. It tends to counteract Calcification in the areas in which they operate together, as they do in soils on the wetter margins of the middle-latitude grasslands. Whereas Calcification tends to concentrate calcium in the lower part of the B

Fig. 3–3. Farm scene in the Blue Grass region near Lexington, Kentucky. The agricultural renown of this region springs, in large part, from the excellence of its soils, which are derived from limestone of high phosphorus content. The most highly prized upland soil of the Blue Grass area belongs to the Maury series, although in early mapping much of it was classed as Hagerstown.

COURTESY OF SOIL CONSERVATION SERVICE, U.S.D.A.

Fig. 3–4. Profile of a well-developed Podzol. Note the distinct light-colored, leached layer (A₂ horizon) about five inches in thickness, capped by a thin layer of undecomposed organic matter and underlain by the dark-colored heavy-textured B horizon. The weathered parent material (C horizon) is seen in the lower one-third of the exposure.

THE PHOTOGRAPH WAS TAKEN IN QUEBEC BY DR. ROY W. SIMONSON AND IS USED WITH HIS PERMISSION.

horizon, Podzolization leaches the entire solum. Not only the calcium bases but other bases as well are removed, and the whole soil becomes distinctly acid. This is the process most active in regions with a cold, humid climate and forest vegetation, but it is not restricted to such an environment. In the lower middle latitudes, and even in the tropics at high altitudes, Podzolization is also active, but there competition with other pedologic processes frequently prevents the completion of the Podzolization process.

Podzolization typifies forest lands in cold climates. Podzolization is chiefly characteristic of the middle latitude and high latitude forests. Its fullest expression is the mature *Podzol* (discussed later) of areas where winters are long, where coniferous forests dominate, where the ground remains frozen for long periods in winter, and all soil-forming activity ceases. Since pedologic processes are inactive in winter, they operate slowly in such cold climates. Consequently, soils take much longer to mature in cold climates than in warm ones, and many cold-climate soils remain submature. In comparison with grasses, trees, particularly conifers, consume only small quantities of bases. They therefore return relatively little basic material to the soil. This shortage of neutralizing bases contributes to soil acidity, which speeds up the Podzolization process.

Podzolization concentrates iron and aluminum. Podzolization concentrates iron and aluminum, and sometimes other materials removed from the upper soil, in the B horizon. Because of this concentration, soils developed largely under the influence of Podzolization formerly were called *pedalfers*. This term was formed by adding "Al" of "aluminum" and "fer" of "ferrum" (iron) to the root of "pedology." The chemical reactions involved in the concentration of these metals are very simple indeed.

The concentration of aluminum in the B horizon contributes to the acidic conditions of the Podzols. Aluminum ions in a water solution hydrolyze ($2Al + 6H_2O \longrightarrow 2Al(OH)_3 + 6H^+$) when they react with water to form both hydrogen ions and aluminum hydroxide ions. (*Hydrolosis* is a process of chemical decomposition in which the ions of a salt react with the ions of water to produce a solution that may be either acid or basic.) Since hydrogen ions are present in abundance, the soil solutions are acid. The presence of iron reacts essentially the same way since the iron reacts with water to form hydrogen ions and ferrous ions. The hydrogen ions make the soil solutions acid as do those of aluminum.

A and B horizons show strong texture contrasts. The B horizon of soils developed under the influence of strong Podzolization has a relatively heavy texture. Above the B horizon the surface soil is coarse textured and strongly leached. There is a whitish-gray layer in the mature Podzol composed largely of finely divided *silica*. This layer, which sometimes is eight to twelve inches thick, forms just beneath the forest litter. A plowed field of Podzol has an ash gray color; the term "Podzol" is derived from the Russian word "Podzol," which means "like ashes."

LATOZATION ACCOMPANIES THE EXTREME WEATHERING AND EXHAUSTION OF SOILS

The Latozation process also operates in regions of heavy rainfall and dense forest vegetation, but unlike Podzolization, it occurs most frequently in warm, humid regions. Soils that develop under this process are badly leached, low in organic and mineral plant nutrients, and very poor for agricultural use.

Since Latozation requires warmth and heavy rainfall, it reaches its peak of development in the very rainy tropics. Here geologic weathering is extremely intense, producing a final product known as the *Laterite*. When the process of Latozation has completed its work, the mature soil produced is known as a *Latosol*. (In Chapter 5, in which the distribution of soils is discussed, the Latosols will be described in some detail.)

Tropical climates stimulate chemical activity. As was discussed in Chapter 2, the high temperatures and abundant rain-

Fig. 3–6. Humic-Gley soils are widespread in areas where drainage is poor, in humid as well as semi-arid climates. This Fillmore silt loam is located in Fillmore County, Nebraska, in a depressed area with very poor drainage. Note the thick, white-Gley layer underneath a dark brown surface layer. The subsoil is a very dark brown silty clay.

COURTESY OF SOIL CONSERVATION SERVICE, U.S.D.A.

fall of the tropics tend to speed up chemical reactions. Warm soil water is plentiful there at all times, and the soluble material is continually being dissolved. When such material is in solution, much of it is carried completely out of the soil by the gravitational water as it moves downward and outward. A warm, rainy climate promotes the continuous operation of the pedologic processes, since there are no cold or dry seasons. Thus soils advance toward maturity and old age much faster than in the middle and high latitudes. Although many tropical areas have luxuriant forest vegetation, the organic remains of such growths are incorporated into the soil very slowly, for much vegetative matter is devoured by ants and other organisms, and, consequently, the soils are surprisingly low in humus. Leaching is pronounced at all times, and the organic matter and soluble minerals are removed rapidly by the acid soil water. Even the silica is leached out. This contrasts sharply with the tendency for Podzols to accumulate a white silica layer. Reddish iron oxide and gray aluminum oxide in strong concentrations are left in both the A and B horizons by Latozation, rather than in the B horizon only, as in the case of Podzolization. The accumulated iron oxide accounts for the strong red color exhibited by many tropical upland soils.

Temperature and moisture conditions set the stage for the active chemical behavior of the Latozation process. The following explanation shows, in a simple way, how some of the reactions take place.

Weathering is very evident in humid sections because of their heavy rainfalls. Iron-containing minerals such as *hornblende, biotite,* and *chlorite* are weathered to produce the iron in the soils. Under ordinary conditions of high rainfall and high temperatures the iron combines with the oxygen in the pore spaces to form a *ferrous* compound, FeO.

$$2Fe \quad + \quad O_2 \quad \longrightarrow \quad 2FeO$$

During soil formation, as the minerals weather, iron may also combine with oxygen to form a *ferric* compound.

$$4Fe \quad + \quad 3O_2 \quad \longrightarrow \quad 2Fe_2O_3$$

The kind of parent material and the conditions of temperatures and rainfall at the time of weathering determine whether the iron compounds exist in the ferrous or ferric state.

Weathering is intense in humid tropical climates, and hydrolysis proceeds rapidly. In the case of hydrated iron oxide ($Fe_2O_3 \cdot H_2O$), the condition is rather unstable, for in a short period of geologic time the iron becomes dehydrated; that is, the water separates from the oxide of iron and the dehydrated Fe_2O_3 is left. Considerable amounts of the iron oxides Fe_2O_3 or FeO stick to the clay, silt, or sand in the soils and form coatings, which may give the soils their color. Some soils have various shades of red, reddish-brown, *ochre,* or yellowish-brown depending upon the presence of various amounts of iron oxides.

As in the case of iron, aluminum is present in the process of Latozation. The aluminum hydrolyzes (combines with water) in a water solution to form aluminum hydroxide and hydrogen ions.

$$2Al \quad + \quad 6H_2O \longrightarrow 2Al(OH)_3 \quad + \quad 3H_2(6H^+)$$

The aluminum oxide adheres to the clay, silt, or sand in the soils to form a gray coating. The shades of gray depend upon the amount of aluminum oxide present.

In summary, the Latosols are high in iron and aluminum but low in silica. It is interesting to note how the action of Latozation is opposed to that of Podzolization in regard to iron, aluminum, and silica. In the Podzols, iron and aluminum leach out leaving only small traces while silica is present in abundance as evidenced by the characteristic white layer. In the Latosols, most of the silica leaches out while iron and aluminum appear plenteously. We must remember, however, that the temperature and rainfall vary considerably. Chemical activities are speeded up a great deal in tropical climates.

FOUR PROCESSES PRODUCE UNADJUSTED INTRAZONAL SOILS

We have examined the three pedologic processes, Calcification, Podzolization, and Latozation, that produce zonal soils. Many soils, however, never reach the stage of full maturity because they lack full adjustment to the genetic factors of environment. These are called *intrazonal* soils. They are discussed in Chapter 4. Most intrazonal soils suffer from both poor drainage and salt accumulation. Intrazonal soils often are found distributed among zonal soils. This is logical, since poor drainage and, to a less extent, salt accumulation are widespread conditions throughout most soil regions.

Intrazonal soils result from the four pedologic processes, Salinization, Solonization, Solodization, and Gleization. They are quite different from the three processes forming zonal soils. Salinization, Solonization, and Solodization operate primarily in subhumid and arid regions. Gleization is chiefly in humid regions.

THREE PROCESSES CAUSE TRANSITIONS IN INTRAZONAL SOILS

The intrazonal soils of dry regions usually are associated areally with zonal grassland soils. Some of them resemble their zonal soil associates; eventually they themselves may become zonal soils after passing through a transitional stage. The three processes, Salinization, Solonization, and Solodization, that form these soils are complicated and are not fully understood. It is known, however, that in many instances these soils are slowly advancing toward a state of equilibrium.

These three processes work in sequence. These pedologic processes seem to serve in bringing about soil transitions by operating in sequence. In other words, one process advances the soil to a certain point in the transition; another process carries it to the next stage, the third to still another stage, and so on until the final result may be the zonal soil common to the area. Thus a particular intrazonal soil that is presently influenced by one pedologic process may have been dominated by one or two other processes in its earlier stages of development. Salinization, Solonization, and Solodization are related in another sense: they all involve the leaching of soluble salts with improvements of natural drainage. Thus it seems better to consider them as a group, rather than separately.

It was pointed out earlier that in soils of subhumid and semi-arid regions, bases tend to accumulate in the profile. Because of an abundance of bases these intrazonal soils are often referred to as "alkali," "saline," or "salty" soils. Light-colored soils are sometimes called "white alkali" soils and dark ones are known as "black alkali" soils. As the pedological processes operate, each at a different stage of the soil transition, they bring about chemical changes, and they also remove much of the soluble basic material.

Each process has a particular role in the transition. Salinization begins the sequence in the genesis of dry region intrazonal soils. In this first stage the soil, known as *Solonchak,* tends to be salty rather than alkaline, since it usually contains an excess of sodium salts. Salinization leaches out much of the salt as drainage improves. Slowly the soil changes from a salty to a strongly alkaline type.

Solonization is the second process to become active, producing a soil known as *Solonetz.* It is dominant while the soil is in the alkaline stage. Leaching continues throughout this stage as natural drainage becomes better. Soils often improve to such a point that they can be used agriculturally. With proper chemical fertilizer, they become quite productive under irrigation and artificial drainage.

Solodization, the third process, follows. It continues to leach out remaining soluble salts as drainage continues to improve. This might be considered the final step in the transition, since the resulting soils, known as *Soloth,* become very much like the zonal soils with which they are associated.

Many intermittent phases may be found between the initial stages of these intrazonal soils and the zonal soils toward which they are developing. Often before Salinization ceases Solonization has already begun. In like manner, Solodization may start while Solonization is still active. Thus, through its profile, a given soil may reveal the effects of two processes operating at the same time.

GLEIZATION FORMS A LIGHT-COLORED GLEY HORIZON

Gleization is the pedologic process responsible for the formation of a light-colored Gley horizon in the mineral part of some soils in poorly drained areas. The Gley layer consists largely of deoxidized iron compounds that accumulate as a result of lack of oxygen. The material is sticky, compact, and structureless; it is grayish or bluish in color. Sometimes the

Fig. 3–6. Landscape scene of Podzol in northern Michigan. Note the cleared tracts of gray Podzols in the distance and a bare spot of the same in the foreground. The coniferous timber dominates in this scene, although mixed hardwoods also thrive in some Podzol areas.

THE PHOTOGRAPH WAS PROVIDED BY DR. ROY W. SIMONSON.

Gley layer is described as "blue mud." Where the ground-water table fluctuates considerably with the seasons, the Gley horizon shows mottling of yellow and rusty brown.

Gleization accompanies organic accumulation. Gleization is characteristic of the accumulation of undecomposed matter (raw humus). In some instances, the organic collection forms a layer lying above the Gley layer. Under these conditions the resulting soil is known by such names as "organic soil," "peat," or "muck". In many instances, however, the organic matter is mixed with the mineral part of the soil.

Gleization thrives where drainage is poor. The Gleization process is dominant in areas of poor drainage, where water-logged conditions prevail much of the time. Intrazonal soils developing under this process are widely distributed over low-land areas, where rainfall is abundant and drainage is poor. The process is not restricted to areas of heavy rainfall, however. It is also active in some subhumid grasslands where drainage is poor. A variety of soils develop under the Gleization process as a result of varying environmental conditions. Some of them are highly productive of shallow-rooted crops when drained, fertilized, and limed. Soil names including *Bog, Half Bog, Humic Gley, Ground-Water Podzol, Muck,* and others are suggestive of the Gleization process.

IN RETROSPECT: SOILS BRIDGE THE GAP BETWEEN THE LIFELESS AND THE LIVING

The reader who searches for a broader and deeper understanding of the world around him now sees soils in their proper perspective with regard to the total physical environment. He sees them as belonging to the living sphere of matter. He sees them as themselves possessing life. He also sees them as performing the all-important function of sustaining plant and animal life throughout the lands of the earth. Finally, he sees soils in the unique role of bridging the gap between the lifeless and the living—between the inanimate rocks of the earth's crust and the complexities of the plant and animal kingdoms.

Soils are evolutionary; they progress in cyclic order. The reader must consider soils in their evolutionary nature, evolving from lifeless rock and advancing in an orderly manner through youth, maturity, and old age, leaving their course clearly marked in their profile features. Little by little he will learn to read the records of natural history that are preserved by their profile features. Further, in order to interpret soils completely, he must also consider them in their full response to the pedologic processes and in relation to the five genetic factors: parent material, plant and animal life, climate, relief, and time.

Soil resources concern all mankind. The unequal distribution of all natural resources, upon which human life depends, is one of the most perplexing problems that confront the human race. Its perplexity increases as the population of the world is multiplied. Since soils are one of our most important natural resources, they are highly prized by peoples who possess them in quality and abundance. Conversely, soils, particularly fertile soils, are jealously coveted by those peoples who hold them in short supply. As a direct source of most food and of many industrial products, the way soils are distributed over the face of the earth is of paramount concern in the economic, social, cultural, and political problems throughout the world.

Word List for Study

weathering	Soloth
humid climate	environment
zonal soils	steppe lands
neutralization	Calcification
Podzolization	saturation
Podzol	pedalfers
pedocals	silica
hydrolysis	Laterite
Latozation	hornblende

Latosol   chlorite
biotite   ferric
ferrous   ochre
hydration   Salinization
intrazonal   Solonization
Solodization   Solonchak
Gleization   Solonetz
                                 Gley

Questions for Consideration

1. Review humus. What is its importance in the soil? How is humus formed?
2. Read about the Grand Canyon of the Colorado. How was it formed?
3. Name and characterize the major processes which form zonal soils.
4. Distinguish between zonal and intrazonal soils.
5. Write the chemical equations for Calcification and explain the part played by the hydrogen ions.
6. Review what is meant by neutralization of soil acidity. How does this take place in the Calcification process?
7. Explain the process of leaching. What is the effect of heavy rainfall?
8. What is the meaning of Pedocal? Pedalfer?
9. Contrast Calcification with Podzolization. How are they different? Are they alike in any respects? Explain.
10. What kind of vegetation may be found in the Podzols? What kind may be found in soils where the process of Calcification is most typical?
11. Explain how the presence of aluminum contributes to the acid condition of the Podzol.
12. Describe the color of the Podzolized soil. Is it recognized easily?

13. Under what conditions does the Latozation process develop the most rapidly?
14. Why do the Latosols leach readily?
15. Distinguish between a ferrous and a ferric oxide. Write equations to explain your answer.
16. Why do the Latosols have a reddish color?
17. What is common to the processes of soil development in intrazonal soils in subhumid and arid regions?
18. Under what conditions may the Solonetz soils become productive agriculturally?
19. Discuss Gleization. What crops can be grown in Gley soils?
20. Explain why soils may be classed as the bridge between the lifeless and the living.

CHAPTER 4

Classes of Soils

SOIL CLASSIFICATION IS BASED ON SOIL CHARACTERISTICS

Soil scientists, like other scientists, must classify the phenomena with which they deal. In fact, classifying the many thousands of individual soils is one of the most important yet most perplexing parts of soil science.

Modern soil classifications are based upon the characteristics of individual soils. Soils might be grouped according to their location, their use, or their market value. They might also be grouped according to some dominant feature, such as color, texture, stoniness, or agricultural productiveness. But such groupings do not tell very much about the characteristics of individual soils. In modern soil classifications, all the properties of soils are considered, and the characteristics that make one different from another are noted.

Emphasis on similarities and differences leads to comparison. In all basic classifications, the similarities and differences of the phenomena are emphasized. The class in which an individual soil is placed depends somewhat upon the ways in which it is like or unlike other soils. Thus classification leads to the comparison of soils with each other and to the grouping of similar soils. When the soils of an area are grouped according to certain given characteristics, soil regions are recognized.

Classification emphasizes the relationships between soils and other natural phenomena. The relationships between soils and other elements of their natural environment often are important in classifying soils and in establishing soil regions. On a worldwide scale, we note that there is a striking relationship between soil regions and climatic regions; on a local scale, we find the relationship of soils to geological conditions best illustrated by the contrast between the soils derived from limestone and those derived from sandstone; while on an intermediate scale, we see the relationship of soils to natural vegetation in the contrast between grassland and forest-land soils.

Different bases for classification have been used. The problem of selecting a basis for classifying soils is one of the oldest problems in soil science. Several different approaches have been followed, each resulting in a different classification. The geological or parent rock material is the basis upon which considerable soil classification is based. The modern classification most commonly employed, however, gives priority to soil characteristics rather than to the materials of which soils are formed. A classification used widely in the United States since the late 1920's had its beginning in the distinction between the soils that accumulate lime and those that do not. (See Chapter 3 on Calcification.)

Obtaining soil data is difficult: the goals are accuracy and usefulness. The problem of obtaining pertinent, reliable data is a major factor in soil classification. Detailed soil surveys are slow and costly. Mapping the soils of a single county may take a field party of several specialists two or three years. Additional work must be done in the laboratory before the soils can be described and classified. When enough relevant data are obtained from field and laboratory work, a long period of study by scores of soil scientists may finally result in a useful classification.

Both accuracy and usefulness are desirable. A given classification may be too technical to be understood except by soil scientists. Another may be too simple for sufficient accuracy. An intermediate position between the very simple and the highly technical extremes is desirable, in that such a classification could be used effectively by intelligent farmers and science students, as well as by soil scientists.

THE DIFFERENT LEVELS OF SOIL CLASSES

Soil science is relatively new. It borrows from several of the older sciences, including geology, physics, chemistry, botany, zoology, and climatology. It makes use of much of the knowledge which has been accumulating over a long period of time. As more is learned about soils, the methods of classifying them change. As classification is refined, it becomes increasingly complex. Only the specialist can be familiar with the details of classification, but many of the broad, general principles can be understood by the student who studies the general pattern of distribution of the major classes of soil.

In classifying soils, as in classifying plants and animals, different breakdowns are necessary to satisfy the various needs that make any classification desirable. Plants might be grouped into forests, grasses, desert shrubs, and tundra for a very general classification. This classification would scarcely satisfy the botanist interested in plant details. In like manner, soils might be grouped in such a way as to depict only the broad classes; for example, the soils of a large world area, such as the hot, rainy tropics or the middle-latitude grasslands. But for specific application, the scientist needs a detailed classification of the soils of individual parts of the area. Thus, any soil classification should entail different levels, or *categories*. The broad levels, or *higher categories,* should be chosen in such a way that they may be subdivided into *lower categories* to meet the need for more details.

THE AMERICAN SOIL CLASSIFICATION AND ITS WORLD-WIDE APPLICATION

One of the classifications now in use by the soil scientists of the United States Department of Agriculture is constantly being refined; yet in general, this classification follows the same broad principles developed about the middle of the twentieth century. It involves several different categories, identified in such a way as to accommodate needs ranging from the broad general interests in soils of large regions to minutely detailed studies of the soils of local areas. The genetic factors of soil formation discussed previously (pp. 34–35) are of basic importance in this classification. The table and discussion that follow present this classification in a somewhat simplified way.

Tables help to present a system. One of the best ways to gain a clear understanding of a classification system is by use of a table or outline. Table 2 [1] is designed to portray the organization of this system. It shows that the classification embraces three higher categories, the *order,* the *suborder,* and the *great soil group.* The lower categories consist of the *family,* the *series,* the *type,* and the *phase.* The lower categories are not included in this table but are introduced later. Each of the higher categories is divided—the order into three parts, the suborder into nine parts, and the great soil group into forty parts (probably about sixty for the entire world.) Each suborder is a subdivision of a specific order and each great soil group is a subdivision of a specific suborder.

From Table 2, we can see that all soils can be traced back to the order, the highest category. Therefore, the student should first consider the three divisions of the order category and relate each one of them to the genetic factors that account for its distinctive role in the classification. The individual suborders suggest the strong influence of a specific environmental factor— temperature, rainfall, drainage, or vegetation. Their function in the classification perhaps is less significant than are those of the order and the great soil group. Except for a highly generalized

---

[1] The two tables in this chapter are basic to understanding soil classification in America, past and present. In studying the terms used in this table it is necessary to associate each of them with its proper category in the system as well as with the kind of soil it represents. The great soil group category is the one with which many people are familiar and the one most used at the regional level. The reader should become familiar with a number of these forty terms.

view of large world regions, or zones, the great soil group and the lower categories provide better approaches than do the order and the suborder.

Soils are classed as zonal, intrazonal, and azonal. A soil classification based on the genetic factors—parent material, vegetation, climate, relief, and time—should consider those soils that are in adjustment with these environmental factors to be norms with which to compare other soils. As was explained in Chapter 3, the soil scientist calls these zonal soils, since they tend to be arranged in zones similar to the broad climatic zones. The world map, page 76, shows seven broad soil zones. It seems clear that many soils may be adjusted to some, but not all, of the genetic factors. The one factor to which soils most frequently are not adjusted is relief, including drainage. Many soils are poorly drained, while others, particularly in dry climates, collect excessive salts. Such soils tend to develop among zonal soils, since drainage problems are factors of most soil regions. They are called intrazonal soils. Still other soils, such as those derived from alluvium and dune sand, never remain stable long enough to develop profile features. They are called *azonal* soils. Zonal soils are the most significant of the three groups, since they represent the normal conditions toward which all soils are progressing.

Great soil groups align with pedologic processes. Within the zonal, intrazonal, and azonal classes many different kinds of soils develop, partly because of differences in the soil-forming processes that bring them about. It was shown previously, for instance, that Calcification develops grass-land soils that contrast sharply with forest-land soils developed under Podzolization, although both are zonal soils. These *great soil groups* make a long list, including the Podzols, the Chernozems, the Red-Yellow Podzolic soils and others found in many places.

The lower categories are the family, the series, the type, and the phase. Within each great soil group are hundreds of soils that differ from each other in their details,

Table 2[1]
Soil Classification in the Higher Categories

ORDER	SUBORDER	GREAT SOIL GROUPS
Zonal soils	1. Soils of the cold zone	Tundra soils
	2. Light-colored soils of arid regions	Desert soils Red Desert soils Sierozem
	3. Dark-colored soils of semiarid and subhumid and humid grasslands	Brown soils Reddish-Brown soils Chestnut soils Reddish Chestnut soils
	4. Soils of the forest-grassland transition	Chernozem soils Prairie soils Reddish Prairie soils
	5. Light-colored Podzolized soils of the timbered regions	Degraded Chernozem Noncalcic Brown or Shantung Brown soils Podzol soils
	6. Lateritic soils of forested warm-temperate and tropical regions	Gray Wooded, or Gray Podzolic soils* Brown Podzolic soils Gray-Brown Podzolic soils Red-Yellow Podzolic soils* Reddish-Brown Lateritic soils* Yellowish-Brown Lateritic soils Laterite soils*
Intrazonal soils	1. Halomorphic (saline and alkali) soils of imperfectly drained arid regions and littoral deposits	Solonchak, or Saline soils Solonetz soils Soloth soils Humic-Glei soils* (Includes Wiesenboden)
	2. Hydromorphic soils of marshes, swamps, seep areas, and flats	Alpine Meadow soils Bog soils Half-Bog soils Low-Humic Glei* soils Planosols Ground-Water Podzol soils
	3. Calcimorphic soils	Ground-Water Laterite soils Brown Forest soils (Braunerde) Rendzina soils
Azonal soils		Lithosols Rigosols (includes Dry Sands) Alluvial Soils

*New or recently modified great soil groups
[1] From the periodical Soil Science, Vol. 67 (1949). p. 118. Authors: Charles E. Kellogg and associates

although all of them have much in common. These individual soils are called soil series. The soils included in a series have essentially the same kind of parent material, the same color, and the same structure and profile features; but differences in texture, slope, drainage, and depth often occur. Soil series are given geographical names taken from localities where they

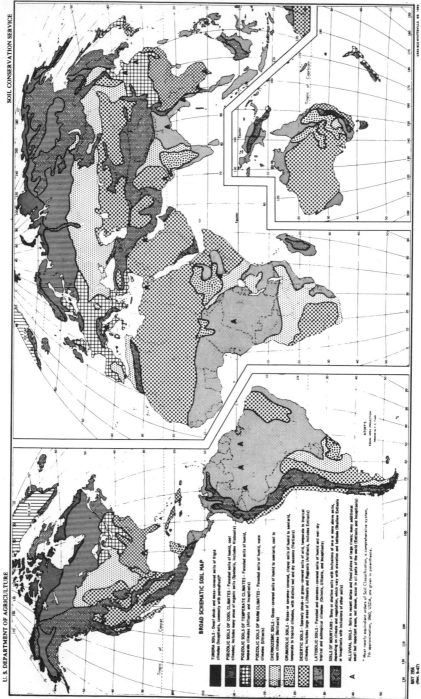

BROAD SCHEMATIC SOIL MAP

TUNDRA SOILS - Dwarf shrub- and moss-covered soils of frigid climates (Inceptisols, commonly with permafrost)*

PODZOLIC SOILS OF COOL CLIMATES - Forested soils of humid, cool climates; includes many areas of organic soils (Spodsols, includes Histosols)

PODZOLIC SOILS OF TEMPERATE CLIMATES - Forested soils of humid, temperate climates (Alfisols and Inceptisols)

PODZOLIC SOILS OF WARM CLIMATES - Forested soils of humid, warm climates (Ultisols)

CHERNOZEMIC SOILS - Grass- or savanna-covered soils of humid to semiarid, temperate to tropical climates, with distinct wet and dry seasons (Vertisols)

GRUMUSOLIC SOILS - Grass- or savanna-covered, clayey soils of humid to semiarid, temperate to tropical climates (Mollisols)

DESERTIC SOILS - Sparsely shrub- or grass-covered soils of arid, temperate to tropical climates; includes large areas of Lithosols and Regosols (Aridisols, includes Entisols)

LATOSOLIC SOILS - Forested and savanna-covered soils of humid and wet-dry tropical and subtropical climates (Oxisols, Ultisols, and Inceptisols)

SOILS OF MOUNTAINS - Stony or shallow soils with inclusions of one or more above soils, depending on climate and vegetation, which vary with elevation and latitude (Shallow Entisols or Inceptisols with inclusions of other soils)

A   ALLUVIAL SOILS - Soils in recent deltas and flood plains of large rivers; many additional small but important areas, not shown, occur in all parts of the world (Entisols and Inceptisols)

* Most nearly equivalent orders of Soil Classification, a comprehensive system, 7th approximation, 1960, USDA, are given in parentheses.

AITOFF'S
EQUAL AREA PROJECTION
Prepared by C. E. Kellogg

Fig. 4-1. Generalized soil map of the world. Note the high degree of generalization, and the tendency of the nine soil groups to conform to climatic and terrain conditions.

were first studied. Norfolk, Hagerstown, Miami, Memphis, and Houston are important soil series. Series often are grouped into soil families, and they may be subdivided into soil types, depending upon the texture: for example, the Norfolk sandy loam is a type of the Norfolk series. The tendency in soil classification, however, is to increase the number of series by subdividing old ones, and to designate fewer types for each series. A soil type may be subdivided into phases, on the bases of such factors as slope, depth, or stoniness. The family, the series, the type, and the phase are called the lower soil categories. They are very important in studying soils on a local scale. A given farm may have several of each. A single county may have forty or fifty series and twice as many types and phases.

Maps focus attention on category levels. The maps in this book illustrate how different categories are used to show the distribution of soils. Figure 4–1 is a highly generalized world map. It shows only seven different classes of soils for the entire world, arranged in zones resembling climatic zones. Several general features stand out (the great deserts, the tundra lands, and the great mountain regions, for example) that permit broad comparisons of large world regions. Figure 4–2 shows the distribution of Great Soil Groups in the United States. The relatively large amount of detail shown on this map gives considerable information concerning soils in different parts of the country and makes possible the recognition and comparison of major soil regions. Obviously, more detail is needed for local areas. Figure 4–3 gives a vast amount of detailed information, depicting the lower categories, the series and even the *type* and the *phase*. This kind of soil mapping is done on aerial photographs, on a county basis, at a very large scale. Often it is necessary to map soils farm by farm, showing minute details that will be useful in planning farm operation and management. Figure 4–5 is a map of an individual farm, which shows not only soils but their recommended use and management as well. Many thousands of farmers in various

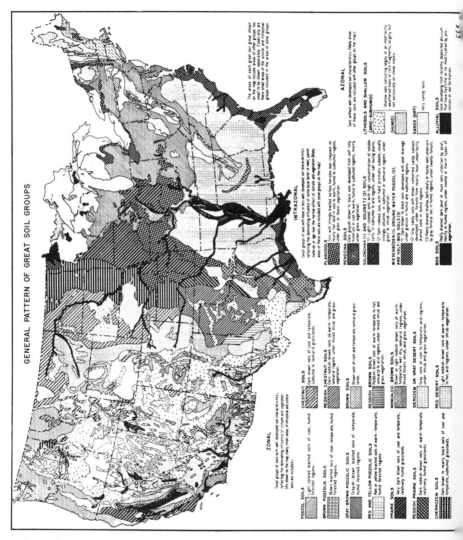

GENERAL PATTERN OF GREAT SOIL GROUPS

ZONAL

Great groups of soils with well-developed soil characteristics, reflecting the dominating influence of climate and vegetation. (As shown on the map, many small areas of intrazonal and azonal soils are included.)

PODZOL SOILS
Light colored leached soils of cool, humid forested regions.

BROWN PODZOLIC SOILS
Brown leached soils of cool-temperate, humid forested regions.

GRAY-BROWN PODZOLIC SOILS
Grayish-brown leached soils of temperate, humid forested regions.

RED AND YELLOW PODZOLIC SOILS
Red or yellow leached soils of warm-temperate, humid forested regions.

PRAIRIE SOILS
Very dark brown soils of cool and temperate, relatively humid grasslands.

REDDISH PRAIRIE SOILS
Dark brown to nearly black soils of warm-temperate, relatively humid grasslands.

CHERNOZEM SOILS
Dark-brown to nearly black soils of cool and temperate, subhumid grasslands.

CHESTNUT SOILS
Dark brown soils of cool and temperate, subhumid to semiarid grasslands.

REDDISH CHESTNUT SOILS
Dark reddish-brown soils of warm-temperate, semiarid regions under mixed shrub and grass vegetation.

BROWN SOILS
Brown soils of cool and temperate, semiarid grasslands.

REDDISH BROWN SOILS
Reddish-brown soils of warm-temperate to hot, semiarid regions, under mixed shrub and grass vegetation.

NONCALCIC BROWN SOILS
Brown or light reddish-brown soils of warm-temperate, wet-dry, semiarid regions, under mixed forest, shrub, and grass vegetation.

SIEROZEM OR GRAY DESERT SOILS
Gray soils of cool to temperate, arid regions, under shrub and grass vegetation.

RED DESERT SOILS
Light reddish-brown soils of warm-temperate to hot, arid regions, under shrub vegetation.

INTRAZONAL

Great groups of soils with more or less well-developed soil characteristics reflecting the dominating influence of some local factor of relief or parent material, or age over the normal effect of climate and vegetation. Many areas of these soils are included with zonal groups on the map.

PLANOSOLS
Soils with strongly leached surface horizons over claypans on nearly flat (and) in cool to warm, humid to subhumid regions, mostly under grass or forest vegetation.

RENDZINA SOILS
Dark grayish-brown to black soils developed from soft limy materials in cool to warm, humid to subhumid regions, under grass vegetation.

SOLONCHAK (1) AND SOLONETZ (2) SOILS
(1) Light-colored soils with high concentration of soluble salts, in subhumid to arid regions, under salt-loving plants.
(2) Dark-colored soils with hard prismatic subsoils, usually strongly alkaline, in subhumid or semiarid regions under grass or shrub vegetation.

WIESENBODEN (1), GROUND WATER PODZOL (2), AND HALF-BOG SOILS (3)
(1) Dark-brown to black soils developed with poor drainage under grasses in humid and subhumid regions.
(2) Light-colored sandy soils with brown-cemented sandy subsoils developed under forests from nearly level, imperfectly drained sand in humid regions.
(3) Poorly drained, shallow, dark peaty or mucky soils underlain by gray mineral soil, in humid regions, under swamp forests.

BOG SOILS
Poorly drained dark peat or muck soils underlain by peat, mostly in humid regions, under swamp or marsh types of vegetation.

AZONAL

Soils without well-developed soil characteristics. (Many areas of these soils are included with other groups on the map)

LITHOSOLS AND SHALLOW SOILS
Shallow soils consisting largely of an imperfectly weathered mass of rock fragments, largely but not exclusively on steep slopes.

(HUMID)

(ARID-SUBHUMID)

SANDS (DRY)
Very sandy soils.

ALLUVIAL SOILS
Soils developing from recently deposited alluvium that have had little or no modification by processes of soil formation.

The areas of each great soil group shown on the map include areas of other groups too small to be shown separately. Especially are there small areas of the azonal and intrazonal groups included in the areas of zonal groups.

Fig. 4–2. Generalized soil map of the Great Soil Groups of the United States (1951)

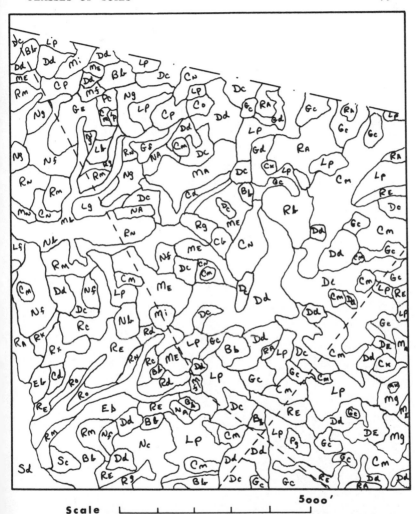

Scale |___|___|___|___|___| 5000'

Fig. 4–3. A highly detailed soils map showing a small portion of Duplin County, North Carolina. The original map from which this was taken is on the large scale of 1:20,000, or 3 1/6 inches to the mile. Note the minute detail, depicting not only the soil series, but the soil type and the soil phase as well.

COURTESY OF SOIL CONSERVATION SERVICE, U.S.D.A.

PATTERNS OF SOIL ORDERS AND SUBORDERS OF THE UNITED STATES

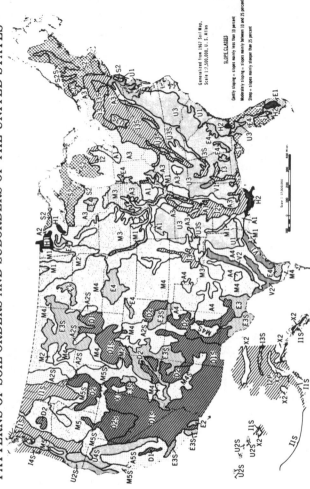

Generalized from 1967 Soil Map,
Scale 1:7,500,000, U. S. Atlas

SLOPE CLASSES

Gently sloping = slopes mainly less than 10 percent
Moderately sloping = slopes mainly between 10 and 25 percent
Steep = slopes mainly steeper than 25 percent

LEGEND

Only the dominant orders and suborders are shown. Each delineation has many inclusions of other kinds of soil. General definitions for the orders and suborders follow. For complete definitions see Soil Survey Staff, Soil Classification, A Comprehensive System, 7th Approximation, Soil Conservation Service, U. S. Department of Agriculture, 1960 (for sale by U. S. Government Printing Office) and the March 1967 supplement (available from Soil Conservation Service, U. S. Department of Agriculture). Approximate equivalents in the modified 1938 soil classification system are indicated for each suborder.

ALFISOLS . . . Soils with gray to brown surface horizons, medium to high base supply, and subsurface horizons of clay accumulation; usually moist but may be dry during warm season

A1 AQUALFS (seasonally saturated with water) gently sloping; general crops if drained, pasture and woodland if undrained (Some Low-Humic Gley soils and Planosols)

A2 BORALFS (cool or cold) gently sloping, mostly woodland, pasture, and some small grain (Gray Wooded soils)

A2S BORALFS steep; mostly woodland

A3 UDALFS (temperate or warm, and moist) gently or moderately sloping, mostly farmed, corn, soybeans, small grain, and pasture (Gray-Brown Podzolic soils)

A4 USTALFS (warm and intermittently dry for long periods) gently or moderately sloping, range, small grain, and irrigated crops (Some Reddish Chestnut and Red-Yellow Podzolic soils)

A5S XERALFS (warm and continuously dry in summer for long periods, moist in winter) gently sloping to steep; mostly range, small grain, and irrigated crops (Noncalcic Brown soils)

ARIDISOLS . . . Soils with pedogenic horizons, low in organic matter, and dry more than 6 months of the year in all horizons

D1 ARGIDS (with horizon of clay accumulation) gently or moderately sloping, mostly used for range (Some Desert, Reddish Desert, Reddish-Brown, and Brown soils and associated Solonetz soils)

D1S ARGIDS gently sloping to steep

D2 ORTHIDS (without horizon of clay accumulation) gently or moderately sloping, mostly range and some irrigated crops (Some Desert, Reddish Desert, Sierozems, and Brown soils, and some Calcisols and Solonchak soils)

D2S ORTHIDS gently sloping to steep

ENTISOLS . . . Soils without pedogenic horizons

E1 AQUENTS (seasonally saturated with water) gently sloping; some grazing

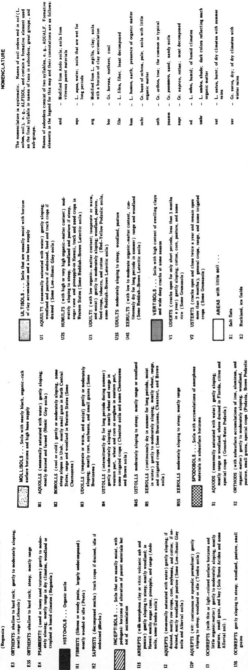

E2 ORTHENTS (loamy or clayey textures) deep to hard rock; gently to moderately sloping; range or irrigated farming (Regosols)

E3 ORTHENTS, shallow to hard rock; gently to moderately sloping; mostly range (Lithosols)

E3S ORTHENTS, shallow to hard rock; steep; mostly range

E4 PSAMMENTS (sand or loamy sand textures) gently to moderately sloping; mostly range in dry climate, woodland or cropland in humid climates (Regosols)

HISTOSOLS ... Organic soils

H1 FIBRISTS (fibrous or woody peats, largely undecomposed) mostly wooded or idle (Peats)

H2 SAPRISTS (decomposed mucks) truck crops if drained, idle if undrained (Mucks)

INCEPTISOLS ... Soils that are usually moist, with pedogenic horizons of alteration of parent materials but not of accumulation

I1S ANDEPTS (with amorphous clay or vitric volcanic ash and pumice) gently sloping to steep; mostly woodland; in Hawaii mostly sugar cane, pineapple, and range (Ando soils, some Tundra soils)

I2 AQUEPTS (seasonally saturated with water) gently sloping, if drained, mostly row crops, corn, soybeans, and cotton; if undrained, mostly woodland or pasture (Some Low-Humic Gley soils and Alluvial soils)

I2P AQUEPTS (with continuous or sporadic permafrost) gently sloping to steep, woodland or idle (Tundra soils)

I3 OCHREPTS (with thin or light-colored surface horizons and little organic matter) gently to moderately sloping, mostly pasture, small grain, and hay (Sols Bruns Acides and some Alluvial soils)

I3S OCHREPTS gently sloping to steep, woodland, pasture, small grains

I4S UMBREPTS (with thick dark-colored surface horizons rich in organic matter) moderately sloping to steep, mostly woodland (Some Brunizems)

MOLLISOLS ... Soils with nearly black, organic-rich surface horizons and high base supply

M1 AQUOLLS (seasonally saturated with water) gently sloping; mostly drained and farmed (Humic Gley soils)

M2 BOROLLS (cool or cold) gently or moderately sloping in North Central States, range and woodland in Western States (Some Chernozems)

M3 UDOLLS (temperate or warm, and moist) gently or moderately sloping; mostly corn, soybeans, and small grains (Some Brunizems)

M4 USTOLLS (intermittently dry for long periods during summer) gently to moderately sloping; mostly wheat and range in western part, wheat and corn or sorghum in eastern part, some irrigated crops (Chestnut soils and some Chernozems and Brown soils)

M4S USTOLLS moderately sloping to steep; mostly range or woodland

M5 XEROLLS (continuously dry in summer for long periods, moist in winter) gently to moderately sloping, mostly wheat, range, and irrigated crops (Some Brunizems, Chestnut, and Brown soils)

M5S XEROLLS moderately sloping to steep; mostly range

SPODOSOLS ... Soils with accumulations of amorphous materials in subsurface horizons

S1 AQUODS (seasonally saturated with water) gently sloping; mostly range or woodland, where drained in Florida, citrus and special crops (Ground-Water Podzols)

S2 ORTHODS (with subsurface accumulations of iron, aluminum, and organic matter) gently to moderately sloping; woodland, pasture, small grains, special crops (Podzols, Brown Podzolic soils)

S2S ORTHODS steep, mostly woodland

ULTISOLS ... Soils that are usually moist with horizon of clay accumulation and low base supply

U1 AQUULTS (seasonally saturated with water) gently sloping; woodland and pasture if undrained, feed and truck crops if drained (Some Low-Humic Gley soils)

U2S HUMULTS (with high or very high organic-matter content) moderately sloping to steep; woodland and pasture if steep, sugar cane and pineapple-in Hawaii, truck and seed crops in Western States (Some Reddish-Brown Lateritic soils)

U3 UDULTS (temperate or warm, and moist) gently to moderately sloping; woodland, pasture, feed crops, tobacco, and cotton (Red-Yellow Podzolic soils, some Reddish-Brown Lateritic soils)

U3S UDULTS moderately sloping to steep; woodland, pasture

U4S XERULTS (with low to moderate organic-matter content, continuously dry for long periods in summer) range and woodland (Some Reddish-Brown Lateritic soils)

VERTISOLS ... Soils with high content of swelling clays and wide deep cracks at some season

V1 UDERTS (cracks open for only short periods, less than 3 months in a year) gently sloping, cotton, corn, pasture, and some rice (Some Grumusols)

V2 USTERTS (cracks open and close twice a year and remain open more than 3 months); general crops, range, and some irrigated crops (Some Grumusols)

AREAS with little soil ...

X1 Salt flats

X2 Rockland, ice fields

NOMENCLATURE

The nomenclature is systematic. Names of soil orders end in sol (L. solum, soil). e.g., ALFISOL, and contain a formative element used as the final syllable in names of taxa in suborders, great groups, and subgroups.

Names of suborders consist of two syllables, e.g., AQUALF. Formative elements in the legend for this map and their connotations are as follows:

and – Modified from Ando soils, soils from vitreous parent materials

aqu – L. aqua, water; soils that are wet for long periods

arg – Modified from L. argilla, clay; soils with a horizon of clay accumulation

bor – Gr. boreas, northern; cool

fibr – L. fibra, fiber; least decomposed

hum – L. humus, earth; presence of organic matter

ochr – Gr. base of ochros, pale; soils with little organic matter

orth – Gr. orthos, true; the common or typical

psamm – Gr. psammos, sand; sandy soils

sapr – Gr. sapros, rotten; most decomposed

ud – L. udus, humid; of humid climates

umbr – L. umbra, shade; dark colors reflecting much organic matter

ust – L. ustus, burnt; of dry climates with summer rains

xer – Gr. xeros, dry; of dry climates with winter rains

Fig. 4-4. This generalized soil map, released in August, 1967, is based on the 7th Approximation classification. It employs the higher categories, Order and Suborder, in establishing soil patterns. This map resembles the 1951 map (shown in Figure 4-2) in some respects, yet they are very different in other respects. The earlier map appears at first to be somewhat more detailed, since it employs 23 patterns of shading, compared with 10 patterns in this map. When the letters and numbers used here to designate soils of specific areas are considered, a total of 41 different soil combinations are found. Thus, the high degree of detail represents a distinct refinement over the earlier map.

COURTESY OF SOIL CONSERVATION SERVICE, U.S.D.A.

parts of the United States are making use of detailed maps of this kind.

There is a category for every need. The person interested in soils from a broad world or a continental point of view can use profitably the world map, page 76. For a single country, such as the United States, the more detailed map showing great soil groups (p. 78) is much more helpful. In studying smaller areas, such as the Kentucky Blue Grass Basin or an individual county, one must work with the lower categories, particularly the series. Published county soil maps provide such details. Soil capability maps for individual farms make detailed use of soil types and phases. Thus, one must choose the soil category best adapted to the scale on which he is working.

A PROPOSED NEW SYSTEM OF CLASSIFICATION

Since the 1950's the soil scientists of the United States Department of Agriculture have been working on a new system of soil classification. They have kept in close touch with soil scientists of other countries, and it is hoped that a universal system adapted to soils of the entire world will result from these efforts. It probably will be several years before this new system is put into wide use outside the United States. It is gaining ground in the United States, however, and may eventually supersede the classification just described. It was officially adopted in 1965 and field work now being done by American soil scientists follows this new system. It should be pointed out, on the other hand, that for several years to come we must rely heavily upon the earlier classifications for literature dealing with individual soils.

The new plan is called: "Soil Classification: A Comprehensive System, 7th Approximation." Its details are complicated and are understandable only to the soil scientist. The system is intended primarily for his use as he works at the job of mapping and classifying soils in the field, in the laboratory, and in the office. One phase of the system, however,

Fig. 4–5. This Conservation Farm Plan illustration shows several different types of land classified according to use capabilities. Such a classification is necessary in constructing a conservation farm plan map such as the one in figures 4–6 and 4–7.

Land Capability Classes: I. Very good land. II. Good land that can be cultivated safely with easily applied practices. III. Moderately good land; requires intensive treatment. IV. Fairly good land; can be cultivated only occasionally. V. (None shown here) Suited for grazing or forestry. VI. Land not suited for cultivation; good for pasture if carefully managed. VII. Land not suited for cultivation; suited for grass or trees with very careful management. VIII. Land not suited for cultivation, grazing, or forestry; good for wildlife or recreation.

COURTESY OF SOIL CONSERVATION SERVICE, U.S.D.A.

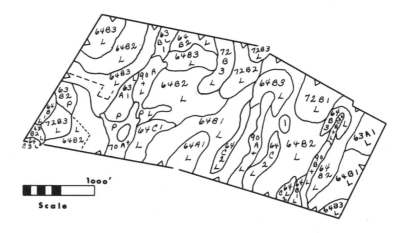

Fig. 4–6. Land capability map of a farm in the Kentucky Blue Grass Basin near Lexington. The land use map of the same farm is shown in Figure 4–7. Figure 4–5 is a landscape view illustrating the way farm land is divided into use capability classes.

COURTESY OF SOIL CONSERVATION SERVICE, FAYETTE COUNTY SOIL CONSERVATION DISTRICT, KENTUCKY.

that will be of interest to all who read the new publications on soils, has to do with the nomenclature, or terms used to refer to soils of different categories.

Old and new ideas merge. In some respects the new system is much like the one that has been in use for a number of years. It approaches the classification problem by using several different categories or levels of soils just as a botanist uses different categories of plants. The highest soils category is called the *order,* the second category is the *suborder,* the third is the *great group,* and the fourth is the *subgroup.* Lower categories include the *family* and the *series.* The nomenclature involves a large number of new terms—terms that have no meaning except as they relate to soils. The higher categories alone have about four hundred such terms.

Each syllable has a meaning; arrangement is important. Without going into the details of this classification and the nomenclature used to express it, a careful look at a few of the terms will illustrate the highly scientific approach employed in coining the long list of technical names. They are formed by combining Latin or Greek syllables with each other or with syllables from other languages, much as terms used in botany, zoology and other natural sciences are formed. Each term is designed to portray a specific fact about soils, and each term can be identified with its proper category by a specific syllable (or part thereof) or combination of syllables.

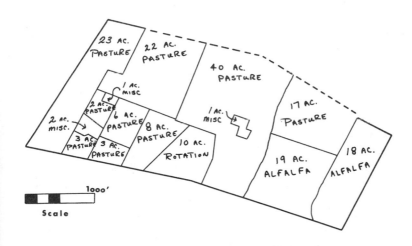

Fig. 4–7. Land use map of the Kentucky Blue Grass farm shown in Figure 4–6. Note the high proportion (71%) of land in pasture, the relatively high significance of alfalfa (21%), and the small proportion of land (8%) in all other uses.

COURTESY OF SOIL CONSERVATION SERVICE, FAYETTE COUNTY SOIL CONSERVATION DISTRICT, KENTUCKY.

For example, ten terms make up the order category (Table 3). Each of these terms ends with the syllable "sol" and it is by this ending that the order category is recognized. "Sol" is preceded by "i" for Latin roots or "o" for Greek roots. About thirty names make up the suborder category. Each of these is a subdivision of one of the ten order names. It is identified by its final part, which is the syllable next preceding the "isol" or "osol" ending of the order name. The third category name, great group (more than one hundred), is formed by prefixing one or more syllables to a suborder name.

Some names require more than one word. Subgroup names consist of the names of the appropriate great group modified by one or more adjectives. Subgroup names, as the term implies, comprise two or more words, such as *Typic Normudolf,* and are subdivisions of great groups. Altogether there are several hundred members of the subgroup category and others may be added as the system develops. Appendix C lists Great Soil Group, Order, Suborder, Great Group, and Subgroup names for 163 American Soil Series. A general subgroup may have all the characteristics that identify the associated great group. On the other hand, it may show its borderline character by having, in addition, some properties associated with another great group. This indicates that a transition from one great group to the other is taking place. In such a case the result is an *intergrade* subgroup. The subgroup category is useful in that it provides a place in the classification for soils in transition, and it offers a means of indicating the direction in which the transition is taking place.

An example illustrates the system. The words "aridisol," "argid," and "durargid" provide an example of the way the classification names are formed and can be interpreted. "Aridisol" (Table 3, No. 4) belongs to the order category; this is evidenced by its *sol* ending. As suggested by the "arid" portion of the word, it embraces the arid, or dry soils. "Argid" is a suborder category indicated by the "id" syllable of "Arid-

Table 3
Names of Orders, Suborders, and Great Groups from the Seventh Approximation

Order	Suborder	Great Group	Order	Suborder	Great Group
Entisols(1)	Aquents	Cryaquents	Inceptisols(3)	Plaggepts	
		Haplaquents			
		Hydraquents		Tropepts	Dystropepts
		Psammaquents			Eutropepts
		Tropaquents			Humitropepts
	Arents				Ustropepts
	Fluvents	Cryofluvents		Umbrepts	Anthrumbrepts
		Torrifluvents			Cryumbrepts
		Tropofluvents			Fragiumbrepts
		Udifluvents			Haplumbrepts
		Ustifluvents			Xerumbrepts
		Xerofluvents	Aridisols(4)	Argids	Durargids
	Orthents	Cryorthents			Haplargids
		Torriorthents			Nadurargids
		Troporthents			Natrargids
		Udorthents			Paleargids
		Ustorthents			
		Xerorthents		Orthids	Calciorthids
	Psamments	Cryopsamments			Camborthids
		Quartzipsamments			Durorthids
		Torripsamments			Paleorthids
		Udipsamments			Salorthids
		Ustipsamments			
		Xeropsamments	Mollisols(5)	Albolls	Argialbolls
Vertisols(2)	Torrerts				Natralbolls
	Uderts	Chromuderts		Aquolls	Argiaquolls
		Pelluderts			Calciaquolls
					Cryaquolls
	Usterts	Chromusterts			Duraquolls
		Pellusterts			Haplaquolls
					Natraquolls
	Xererts	Chromoxererts			
		Pelloxererts		Borolls	Argiborolls
Inceptisols(3)	Andepts	Cryandepts			Calciborolls
		Durandepts			Cryoborolls
		Dystrandepts			Haploborolls
		Eutrandepts			Natriborolls
		Hydrandepts			Paleborolls
		Vitrandepts			Vermiborolls
	Aquepts	Andaquepts			
		Cryaquepts		Rendolls	
		Fragiaquepts			
		Halaquepts		Udolls	Argiudolls
		Haplaquepts			Hapludolls
		Humaquepts			Paleudolls
		Plinthaquepts			Vermudolls
		Tropaquepts			
	Ochrepts	Cryochrepts		Ustolls	Argiustolls
		Durochrepts			Calciustolls
		Dystrochrepts			Durustolls
		Eutrochrepts			Haplustolls
		Fragiochrepts			Natrustolls
		Ustochrepts			Paleustolls
		Xerochrepts			Vermustolls

Table 3 (Cont.)
Names of Orders, Suborders, and Great Groups
from the Seventh Approximation (Cont.)

ORDER	SUBORDER	GREAT GROUP	ORDER	SUBORDER	GREAT GROUP
Mollisols(5)	Xerolls	Argixerolls			
		Calcixerolls	Alfisols(7)	Xeralfs	Durixeralfs
		Durixerolls			Haploxeralfs
		Haploxerolls			Natrixeralfs
		Natrixerolls			Palexeralfs
		Palexerolls			Plinthoxeralfs
					Rhodoxeralfs
Spodosols(6)	Aquods	Cryaquods			
		Duraquods	Ultisols(8)	Aquults	Fragiaquults
		Fragiaquods			Ochraquults
		Haplaquods			Plinthaquults
		Placaquods			Tropaquults
		Sideraquods			Umbraquults
		Tropaquods			
				Humults	Haplohumults
	Ferrods				Palehumults
					Tropohumults
	Humods	Cryohumods			
		Fragihumods		Udults	Fragiudults
		Haplohumods			Hapludults
		Placohumods			Paleudults
		Tropohumods			Plinthudults
					Rhodudults
	Orthods	Cryorthods			Tropudults
		Fragiorthods			
		Haplorthods		Ustults	Haplustults
		Placorthods			Paleustults
					Plinthustults
Alfisols(7)	Aqualfs	Albaqualfs			Rhodustults
		Fragiaqualfs			Tropustults
		Glossaqualfs			
		Natraqualfs		Xerults	Haploxerults
		Ochraqualfs			Palexerults
		Tropaqualfs			
		Umbraqualfs	Oxisols(9)	Aquox	Gibbsiaquox
					Ochraquox
	Boralfs	Cryoboralfs			Plinthaquox
		Eutroboralfs			Umbraquox
		Fragiboralfs			
		Glossoboralfs		Humox	Acrohumox
		Natriboralfs			Gibbsihumox
		Paleboralfs			Haplohumox
					Sombrihumox
	Udalfs	Agrudalfs			
		Ferrudalfs		Orthox	Acrorthox
		Fragiudalfs			Eutrorthox
		Glossudalfs			Gibbsiorthox
		Hapludalfs			Haplorthox
		Natrudalfs			Umbriorthox
		Paleudalfs			
		Tropudalfs		Torrox	
	Ustalfs	Durustalfs		Ustox	Acrustox
		Haplustalfs			Eutrustox
		Natrustalfs			Haplustox
		Paleustalfs			
		Plinthustalfs	Histosols(10)	Incomplete	
		Rhodustalfs			

1. From **Soil Classification—A Comprehensive System, 7th Approximation** Supplement, March, 1967.   Soil Conservation Service, U. S. Department of Agriculture.

isol." "Arg" comes from the Latin word "argilla," meaning white clay. So, the "Argid" suborder consists of arid soils with a white clay horizon. "Dur" (Latin: "durus," durable, hard) may be prefixed, making the three-syllable great-group name "Dur-arg-id," meaning arid soils with a hard white clay horizon.

Table 3 fits only the higher categories. Table 3 provides a framework for grouping the higher soil categories only. It illustrates how categories can be combined to form progressively higher levels. For example, each member of the great group category finds its place in the suborder level, and finally in the highest category, the order. Thus, each great group belongs to a specific suborder, which in turn, belongs to a specific order.

Table 3, however, is not concerned with the lower soil categories, and no such table has yet been developed. Members of the lower categories, the family and the series, combine to make up the subgroups and the great groups in much the same way that the great groups make up the suborder and, finally, the order categories.

The series is an old and little changed category. The series is the lowest category in this new classification. The series name usually is taken from the geographical location where the soil was first identified and studied. Thus the series was established as a category when soil classification first began in America. Miami, Gloucester, Cecil, and Hagerstown are familiar series names. In this new classification, the series are identified and described by studying very small soil units, called *pedons,* which usually are less than 10 square meters in area. There are more than eight thousand soil series in the United States alone, and the list is continually lengthening.

The family is the least acceptable category. The soil family is a combination of related series. No definite system has yet been devised and field-tested whereby the family category can be fitted smoothly and effectively into this new classification. The tendency has been, and still is, to give the family the name of its dominant series. Other methods of establishing and naming soil families are under consideration. Criteria have not been definitely selected and adequately field-tested; thus, the soil family as a category in the new classification awaits further study and refinement.

The new approach is a forward step, but progresses slowly. The proposed approach to soil classification is an important step forward in soil science. By means of the new approach, soil science may be organized in much the same way that other sciences are organized. Soils will be presented as they are actually found to be when studied in place, not as they are expected to be because of relationships they might have with other environmental factors, or because of influences of the environment of the past. The new system will give soil science its own nomenclature of scientifically coined names that can be associated with technical terms used in other sciences. Above all, it will lead to greater precision in identification and interpretation of soils.

The new classification has not progressed far enough to permit its use in presenting a world picture of soil distribution. It has not yet been fully proven except on a local scale. When it does gain wide acceptance, since it is designed for the soil scientist and not for the beginner, it might well prove quite difficult to present.

Word List for Study

soil classes	suborder
categories	family
order	type
Great Soil Group	azonal

series

tundra

phase

great group

intrazonal

subgroup

soil series

Sierozem

Chernozem

Solonchak

Rendzina

alluvium

profile

Podzol

arid

pedons

Questions for Consideration

1. What is the broad basis of modern soil classifications?
2. What relationships are considered in determining the class into which a soil is placed?
3. Review the world map on page 76. What similarities can you see between climatic zones and soil regions?
4. Distinguish between lime-accumulating soils and those which do not show appreciable amounts of lime.
5. What is the difference between higher categories and lower categories of soil classification?
6. How are the genetic factors of soil formation related to the classification of soils most commonly used today?
7. Study Table 2, p. 75. What are the main differences between zonal and intrazonal soils?
8. Why is it that soils developed from dune sand and river alluvium do not develop profile features?
9. Review Calcification and Podzolization. How are the

grasslands of the Calcification process similar to the forest lands of the Podzolized soils?

10. What is a soil series? How do the soils of a series resemble one another and how may they differ?

11. How do the soil series get their names?

12. What is a soil capability map? How may farmers use these maps beneficially?

13. Give some specific examples of how Latin and Greek are used in the new system of soil classification.

14. What is the meaning of the word *pedon*? How does it differ from the soil series?

15. Read about Curtis Marbut in a good reference book or an encyclopedia. Why would you call him a true soil scientist?

CHAPTER 5

Distribution of Soils

SOIL CLASSES FIT GEOGRAPHICAL REGIONS

The preceding chapters of this book are concerned with the nature of soils, the factors and processes of soil formation, and the classes into which individual soils are grouped. This chapter deals with the distribution of soils. We have seen that different soils intermingle to varying degrees in areas of all sizes. Rarely is a single kind of soil spread widely over a large area. A given farm may have five to ten different soil series, and two or three great soil groups may be represented in it. An average-sized county often has a score or more series and three or four great soil groups.

The complex arrangement of soils in nature makes it difficult to describe their distribution over the lands of the earth. One might concentrate on a particular kind or class of soil and describe its worldwide distribution. To do this for all soil classes would begin and end in confusion. In like manner, a single quality or combination of qualities—color, texture, or drainage—will prove unsatisfactory for showing soil distribution.

Climate, vegetation, and relief help set soil patterns. The close relationship between soils and other elements of their environment suggests an appropriate approach to soil distribution. Broad natural regions marked by climate, vegetation, and relief perhaps provide the best framework for presenting the world distribution of soils. In this chapter we

Fig. 5–1. This attractive Virginia farm scene shows results of scientific methods and good management. In spite of unfavorable slope and shallow, cherty soils, the farm displays a distinct air of prosperity and charm. Soils belong largely to the Clarksville and Lodi series, which never enjoy higher than moderate rating for agriculture.

COURTESY OF TENNESSEE VALLEY AUTHORITY.

recognize five broad subdivisions of the lands of the world for the purpose of showing soil distribution. These are:

1. soils of the humid lands
2. soils of subhumid grasslands and deserts
3. soils of tundra regions
4. alluvial soils
5. mountain soils.

The highest categories fit only broad, general patterns. The American Soil Classification, discussed in Chapter

Fig. 5–2. This rolling Midwestern landscape shows several different soil series, each devoted to a different use. In the foreground is corn on nearly level Zanesville-Tilset soils, adjoining steeper Welston-Zanesville soil planted in wheat. Wet alluvial Stendal soils are in permanent pasture. Timbered soils in the upper left corner belong to the Welston Series.

COURTESY OF SOIL CONSERVATION SERVICE, U.S.D.A.

4, has been tested and adapted to many foreign as well as American areas. Therefore it has been selected in this study to show world soil·distribution. Because our purpose is to present a generalized picture, only the higher categories can be used effectively.

Of the larger world subdivisions listed above, the names of the first three indicate a strong climatic influence. Thus, since the areas resemble broad climatic zones, it appears that the zonal category may apply to them. The other two subdivisions, alluvial and mountain soils, indicate influences other than climate. This suggests, then, that intrazonal and azonal categories are dominant. Within each major subdivision the great soil group category helps present the details of distribution patterns.

The lower soil categories are suited to local patterns. The series, the type, and the phase comprise the lower soil categories. They fit well into the distribution patterns of small areas. This is particularly true where local soils contrast strongly with each other in appearance, productivity, and other characteristics. Two examples will illustrate this point:

1. The bright red Cecil clay loam of the Appalachian Piedmont contrasts sharply with the gray Durham sandy loam beside it.

2. The highly productive brown Hagerstown loam (from limestone) outproduces, by far, its neighboring sandy Muskingum (from sandstone) in the Appalachian Valley area.

Careful interpretation of detailed soil patterns can contribute much to the analysis of local regions.

SOILS OF THE HUMID LANDS—FORESTS

Most humid lands of the world are forested, and conversely, most forests are located in areas of relatively abundant precipitation. Some forests, however, are in cold areas where precipitation, though relatively light, is highly efficient. Forests vary greatly in character and, as a result, forest soils likewise vary. In the broadleaf evergreen forests of the tropics there is relatively little return of vegetative matter to provide abundant soluble bases, although the forest vegetation in most places is luxuriant. Since there is no winter and no season of drought, trees do not shed their leaves regularly; hence, little leaf mold collects. Most plants are perennials, rooting rather deeply. There is little annual return of dead plants and decayed roots. The fallen bodies of dead trees are quickly devoured by myriads of ants and other insects. The hot, rainy climate is conducive to rapid chemical reaction and excessive leaching of soluble soil materials. For these reasons, wet tropical soils are low in organic material, greatly leached, and among the world's poorest agricultural soils.

Middle latitude forest soils are somewhat better off than soils of the tropical rain forests. Lower temperatures retard soil leaching. The cold of winter brings about a dormant season for deciduous hardwood species, returning to the ground a leaf litter with each passing year. In the near-absence of devouring insects, the leaf mold gradually becomes humus with continuous, but relatively slow decay. Evergreen conifers provide a tremendous volume of litter, exceeding in annual weight the leaf fall of most deciduous hardwoods. Pine needles often accumulate to depths of several inches. Yet, soils under coniferous forests are less favored with organic material than soils of deciduous forests. Evergreen conifers do not shed their needles regularly, and the litter is incorporated into the soil more slowly and less effectively than in the case of deciduous trees. Furthermore, pine needles, as well as needles from some other conifers, tend to increase the soil acidity, thereby lowering its quality for most agricultural crops. In the case of all types of forests—tropical evergreen, temperate deciduous, and temperate coniferous—it is worth noting that most of the organic material that accumulates is left *on* the surface, rather than *in* the soil, as in the case of grasslands.

Humid-land soils are badly leached. Soils of the humid lands represent two extreme classes, with transitions between. Two processes, Podzolization (most active in cold, humid lands), and Latozation (most active in warm, humid lands), tend toward extreme leaching of soluble bases and other plant nutrients. Although they work in very different ways, each process, when not interfered with, produces a soil of very poor agricultural quality. In the transition areas where both processes work in competition the resulting soils are much better than either extreme class. Regardless of intense leaching and generally poor agricultural qualities, humid-land soils are highly important to world civilization. Most of the world's people live in humid lands, obtaining the bulk of their food and many other products from these relatively poor soils.

PODZOLS TYPIFY COLD FORESTS

Podzols (see p. 56) are the typical mature soils of regions

having humid subarctic climate, although they dominate some areas with humid continental climate as well. Not only are they poor for agriculture but they occupy, for the most part, areas that have poor agricultural climates. Long, cold winters and cool, short summers restrict the choice of crops to the most hardy ones. At present, the agricultural use of Podzols, particularly in America, is not very widespread. Early attempts to farm them, especially the thin, stony soils of northern New England and eastern Canada, were only partially successful. Strong acidity and low plant-nutrient content require heavy liming and fertilizing. Careful handling to make farming profitable is necessary on many Podzols. Those derived from calcareous bedrock or from calcareous glacial till are less leached, somewhat better supplied with plant nutrients, and finer-textured than Podzols formed from noncalcareous materials. Potatoes and other acid-tolerant root crops—rye, oats, and grasses—yield fair returns on the better Podzols. Dairying is one of the more profitable types of farming in the more densely settled Podzol areas.

As with all the great soil groups, mature Podzols comprise only part of the soils of Podzol areas. Mixed with zonal soils are poorly drained Groundwater Podzols, peat and muck lands, river alluvium, and other intrazonal and azonal soils.

Good forests grow in Podzols. Podzols are well suited to coniferous trees, to some deciduous trees, and to some other acid-tolerant plants. They produced much of our early supplies of softwoods. Maine, the Pine Tree State, and her New England neighbors were unmatched by other forest areas known in colonial times as a source of ship masts and other structural timbers. A large proportion of Podzol soils still is given over to forest growth, much of which has been cut over one or more times.

PODZOLIC SOILS ARE NOT THOROUGHLY PODZOLIZED

Southward from the Podzol regions the mature soils also have developed largely under the influence of Podzolization; but they are not Podzols. They consist primarily of the Gray-Brown Podzolic and the Brown Podzolic groups. In America they occupy a belt from southern New England to Maryland, and westward to the prairies of the Middle West. From the standpoint of soil development this area differs from the Podzol environment in two important ways: (1) the climate is milder, with longer, warmer summers and shorter, less severe winters, and (2) the natural vegetation consists primarily of deciduous forests with an understory of shrubs, herbs, and

Fig. 5–3. The Miami silt loam pictured here is widely distributed in Ohio, Indiana, and other Midwestern states. It is one of the better Gray-Brown Podzolic soils of the area. Because of 3% slope and considerable erosion, the soil pictured here is classed below average for the Miami series.

COURTESY OF SOIL CONSERVATION SERVICE, U.S.D.A.

grasses, sometimes mixed with conifers. These Podzolic soils, in comparison with true Podzols, have a higher content of bases and organic matter, and are less leached because the hardwood trees, along with the understory, return each year a crop of leaves relatively rich in bases.

Podzolic soils have many good qualities. The less acid condition of the Podzolic soils favors the existence of worms, bacteria, and other organisms that assist in incorporating humus into the surface soil, giving it a darker color than the characteristic gray layer of the Podzol. The B horizon is somewhat heavier than the A, due to the accumulation of iron, aluminum, and clay. The structure usually is better than in

Fig. 5–4. **This Profile of Norfolk sand taken in eastern North Carolina** shows a thick A horizon of fine sand underlain by a yellowish sandy loam B horizon. Norfolk sand, which is widely found in the Atlantic Coastal Plain, is inferior to the sandy loam type for most crops.

PHOTO BY J. SULLIVAN GIBSON.

Podzols, the nut type being most common. The higher content of bases and other plant nutrients, good circulation of water and air, and free movement of plant roots help make Podzolic soils responsive to fertilization.

Many soils not belonging to the mature Podzolic groups are mixed with them. Poorly drained intrazonal soils, river alluvium, and thin, stony soils such as those of the steep Appalachian slopes dominate in local areas. Even the mature zonal soils are not all productive agriculturally; and forest stands, several times cut over, still predominate over agricultural lands in some areas.

Podzolic soils favor a variety of crops. Most crops and farming practices brought to America by early settlers were first developed on similar European soils. Little experimentation was necessary. From the early beginning through a long period of expansion as far west as the upper Mississippi River, agriculture was successful on these soils. Podzolic soils are equally well adapted to the subsistence agriculture of earlier times and the commercial agriculture of today. They respond well to good management and fertilization. A wide variety of crops—including all important grains (except rice), forage crops, tobacco, root crops, many truck crops, and orchard fruits—along with dairying and livestock farming, thrive in them under proper cultivation. Under good management the Podzolic soils are among our best forest soils for general agriculture, involving a variety of crops and livestock.

LATOSOLS ARE TYPICAL TROPICAL SOILS

Opposite to the Podzols are the Latosols of the tropics, the final result of the Latozation process. Tropical soils and Latozation are not as well understood as are soils of the middle and high latitudes and Podzolization. Latozation is so like Laterization, a geologic process of weathering, that the two are often confused.

Typical tropical soils develop under evergreen forests, although some develop under grasses. In either case the soils are greatly leached of bases and organic matter; they are also low in silica, high in oxides of iron and aluminum, and have a relatively high clay content.

Latosols are the world's most weathered soils. Latosols are the world's most strongly weathered soils, but in spite of deep weathering, the profile features are sometimes not distinctly marked. The B horizon, with a concentration of iron and aluminum oxides, frequently exhibits hardpan tendencies. Colors vary, but red and yellow hues from the iron and aluminum oxides are most common. Many tropical soils are granular, porous, and very permeable to water.

Good trees, not good crops, grow in Latosols. Tropical soils in general are very poor for agricultural crops. This is true because the luxuriant forest growth usually has trees which are deep feeders. As a result, relatively limited nutrient supplies are left for shallow-feeding annuals once the forest is cleared. Many people in tropical lands commonly practice a migratory system of farming involving frequent abandonment of fields for fresh, new clearings. Where plantation agriculture is practiced, extreme care in selecting soils is necessary.

Included with the zonal Latosols are poorly drained bog and other intrazonal soils, river alluvium, porous sands, and volcanic ash.

PODZOLIC-LATOZOLIC SOILS CHARACTERIZE THE SUBTROPICS

Podzolic-Latozolic soils embodying some of the characteristics of both Podzolization and Latozation occur in many parts of the tropics and subtropics. These transitional soils are found in areas of considerable size and of great agricultural significance in southeastern United States and southern China.

Latozation increases southward. The Red-Yellow Podzolic soils of the United States have something in common with the Gray-Brown Podzolic soils further north. 'Podzoliza-

Fig. 5–5. Plowed field of Norfolk fine sandy loam. This excellent soil for tobacco, peanuts, and cotton is widely distributed over the inner Coastal Plain from Chesapeake Bay to the Gulf of Mexico.

COURTESY OF SOIL CONSERVATION SERVICE, U.S.D.A.

tion has been an important process in the formation of each; but in the Red-Yellow group, competition with Latozation has weakened the effects of Podzolization. Along their common boundary, from Virginia westward to Oklahoma, the Podzols and Latosols are much alike. To the south the influence of Latozation increases, as evidenced by the strengthening of the red and yellow colors and an increase in leaching. The soils are deeply weathered and susceptible to erosion.

Red-Yellow Podzolic soils are excellent for cotton. Red and Yellow Podzolic soils are the typical Cotton Belt soils of the South. Each is named after the characteristic color of its subsoil. The Yellow Podzolic soils occupy the more level lands of the Coastal Plain and the smooth uplands of the Piedmont

Fig. 5–6. This profile of Oktibbeha loam is seen near Montgomery, Alabama. The white limy layer at the bottom of the cut is Selma chalk but the overlying layer is derived from the geologically younger Ripley formation. The Oktibbeha, together with the Houston and the Sumter series, are the famed cotton soils of the Alabama Black Belt.

PHOTO BY JULIAN PETTY.

and the Appalachian Plateau. The native vegetation on these soils was largely pine but with an admixture of hardwoods in places. Red Podzolic soils occupy the undulating, sloping, and hilly lands. Deciduous hardwoods originally were dominant in the vegetation, with an occasional mixture of pines. Good drainage of these soils is thought by some soil scientists to be partly responsible for their red color, whereas the yellow soils may owe their color to inferior drainage conditions during their development.

Red-Yellow Podzolic soils respond well to fertilization and good treatment. Red-Yellow Podzolic soils in their natural state are relatively poor for agriculture. They are rather strongly acid, low in organic matter, and poor in most plant nutrients. Most of the better soils were brought under cultiva-

tion early, and continuous cropping in cotton, corn, and tobacco resulted in accelerated erosion. This, together with
exhaustion of nutrients, has resulted in repeated abandonment
of many fields. Nevertheless, abundant rainfall and a long
growing season encourage agricultural efforts. For more than a
generation the region has been a heavy user of commercial
fertilizer, to which these soils respond unusually well. After
the decline of cotton growing in the South and Southwest,
diversification began rapidly to replace the one-crop system
there. As a result, there has been a great improvement in the

Fig. 5–7. This field of Appling fine sandy loam is located near Raleigh,
North Carolina. The Appling series is a light red Piedmont soil derived
from granitic rocks. It is a general purpose soil and is widely distributed.
Evidence of erosion can be seen in the foreground.

Fig. 5–8. This is a profile of Porters loam, a Gray-Brown Podzolic soil found widely in the Appalachians. Note the dark surface layer, high in organic matter, and the moderate depth to rock.

programs of soil management in the states where cotton once was king.

Excellent timber grows in the Red-Yellow Podzolic soils of the South. The Red-Yellow Podzolic soils of the southeastern states have long been an important source of both pine and hardwood timber. The Southeast is still one of the nation's chief sources of lumber and pulpwood. Both the soils and the climate in this region favor the rapid growth of trees. Some of the soils there, particularly the most severely eroded ones, are more valuable for timber than for crops.

As in most soil groups, many intrazonal and azonal soils are interspersed among the zonal soils of the Podzolic-Latozolic group. For example, the rivers have deposited alluvium over large areas of these soils; again, there are extensive tracts of poorly drained soils in the bogs and swamps, particularly in the tidelands and the interstream swamps of the Coastal Plain.

SOILS OF THE SUBHUMID GRASSLANDS AND DESERTS

It has been stated earlier that grassland and desert soils are formed primarily under the influence of the Calcification process. Since this is the dominant process in climates ranging from subhumid to arid, the resulting soils vary greatly. As we move from subhumid to semiarid to arid areas, changes in soils take place gradually. Less moisture means less leaching and more Calcification; it also means less vegetation, less humus, and lighter color. These grassland and desert soils are made up of three broad, general classes: (1) Chernozemic-grassland soils appear in areas of heavy rainfall, (2) Chernozemic-desertic soils occupy intermediate rainfall areas, and (3) Desertic soils predominate in the dry extremes.

CHERNOZEMIC SOILS ARE WIDESPREAD

The Chernozemic class includes several well-recognized zonal soil groups, of which the Brunizem (formerly called Prairie) and the Chernozem groups are most significant.

Brunizem soils occupy the rainier grasslands. Brunizem soils represent the rainier grassland regions of the middle latitudes. Tall grasses comprise the native vegetation, yet the rainfall is sufficient to prevent a lime layer from forming.

Leaching goes on slowly while Calcification is in progress, indicating competition between the Calcification and the Podzolization processes. Brunizem soils are neutral to moderately acid in their reaction. Deep and abundant accumulation of organic matter derived from grass roots accounts for the dark brown to black color and in part for fine granular texture of both surface and subsoil horizons. Large areas of Brunizem soils are located in the United States, The Soviet Union, and South America. The map, page 78, shows a large area of Brunizem soils in the Corn Belt states, which extends from western Indiana to western Iowa.

Brunizem soils are world-renowned producers. Brunizem soils are among the world's most productive soils. They are especially well adapted to a variety of crops including corn, small grains, soy beans, hay, and pasture grasses. The qualities of good texture and structure and the abundant soil moisture associated with them, make them easy to cultivate. Typical zonal soils occupy rolling divides, where the drainage usually is adequate, but under less favorable drainage conditions tile drainage is practiced. Brunizem soils are susceptible to erosion, and where slopes are steep they require careful handling. Light applications of lime and fertilizer make these Brunizem soils highly productive.

Rendzinas are not true Brunizem soils. An intrazonal group of grassland soils called Rendzinas are very much like Brunizem soils in many respects, although usually the two groups are not associated geographically. Rendzinas develop from soft, highly calcareous material, such as marl, which disintegrates rapidly, maintaining high lime content in the soil. They are found frequently in rainy climates adapted to forest growth, yet the natural vegetation is predominantly grass. Soils of the Black Prairies of Texas and the Black Belt of Alabama are usually classed as Rendzinas, although they are more accurately called "Grumusols."

Mixed with zonal Brunizem soils are poorly drained intra-zonal soils and river alluvium. On their rainier forested borders Podzolic and Brunizem soils dovetail.

Chernozems are the darkest grassland soils. Cherno-zems (meaning "black earth" in Russian) are the most fertile of the grassland soils, although they are not the most produc-tive, since they occupy areas of light rainfall. Chernozems are formed under dense covers of grasses, some tall and some short. They owe their dark-brown to jet-black color largely to their very heavy charge of organic matter, derived chiefly from grass roots. Chernozems have a well-flocculated structure and a relatively large component of clay, which makes them very

Fig. 5–9. Profile of Crete silt loam, a Chernozem found in Republic County, Kansas. Note the dark surface soil with a heavy charge of organic matter, characteristic of many grassland soils. The lime layers are clearly shown below the solum.

COURTESY OF SOIL CONSERVATION SERVICE, U.S.D.A.

sticky when wet and susceptible to erosion. The true Chernozem has a layer of lime, usually within 3 to 5 feet of the
surface, in easy reach of the deep-rooted grasses. Reserves of
plant minerals and organic matter are so abundant that successive cropping without fertilizing is possible over long periods
of years.

Chernozems are superior wheat soils. Chernozems are
characteristic of plains areas with approximately 20 inches of
annual precipitation. In America and in the Soviet Union they
stretch from the drier margins of the Brunizem regions to the
borders of the Chernozemic-desertic soils. The Dakotas and
eastern Nebraska represent the heart of the American Chernozem regions. Chernozems are unusually good grain lands, pro-

Fig. 5–10. Landscape view of a Chestnut soil area in Kansas. This is
typical wheat and corn country. Note: (1) level terrain (2) rectangular
pattern, and (3) complete utilization with practically no waste land.
COURTESY OF SOIL CONSERVATION SERVICE, U.S.D.A.

ducing a large proportion of the world's wheat. Most of the Chernozemic land that is not in cultivation is used for pasture.

Chernozem boundaries are rather indistinct in most places, the zonal soils merging with Brunizem soils on the wetter margins and with Chestnut soils on the drier borders. Intrazonal soils mixed with Chernozems belong largely to arid-climate groups that have acquired excessive salts or alkalies.

CHERNOZEMIC-DESERTIC SOILS ARE FAMED FOR GRAZING

On the drier margins of the Chernozems are soils belonging to the Chernozemic-desertic class. These soil regions have long been important grazing lands in all the inhabited continents (see map, p. 76). They have developed under semiarid conditions, with annual precipitation ranging from 20 inches to as low as 10 to 12 inches on desert margins. Vegetation consists of moderate to light stands of short grass, mixed with desert shrubs in drier areas. Chernozemic-desertic soils show strong Calcification influence in their low degree of leaching and in the lime layer, usually within 1 to 3 feet of the surface. Since rainfall is lighter and vegetation is sparser than with the Chernozems, these soils are much lower in organic matter and lighter in color.

Chernozemic-desertic soils are brown or reddish. In North America, Chernozemic-desertic soils extend in a belt through the Great Plains from southern Canada to central-western Texas. Several zonal groups are represented. Chestnut soils border the Chernozems in the northern part of the region. They are the darkest. Lighter-colored Brown soils join with the Chestnut group on their drier margin. Further south both the Chestnut and the Brown soils become reddish, as a result of greater oxidation associated with higher temperatures.

Numerous intrazonal soils with alkali and salt accumulations are included with the Chernozemic-desertic soils. Also, numerous sand areas, of which the Sand Hills region of Nebraska is an example, are classed as azonal soils.

Farming gives way to ranching in drier parts. Precipitation is too light and uncertain for crop raising except in the rainier portions of these soil regions. Where irrigation is practiced, usually on a local scale, these soils produce heavy yields. Wheat is the leading crop in the northern part of the American region. From Kansas southward, grain sorghums compete with wheat. The Texas Panhandle area has developed an important cotton crop, but wheat and grain sorghums also are important there. Since per-acre yields are low, farms are necessarily large throughout the region. In all areas of Chernozemic-desertic soils pasturing is a major activity. In some drier areas ranching prevails almost to the exclusion of crop raising.

DESERT SOILS ARE HIGH IN BASES

Desert soils develop under extreme conditions of aridity. The accumulation of basic materials results in deposits of lime and other whitish mineral substances being left near to, or even upon, the soil surface. Desert soils in general are light in color, partly because of this accumulation of light mineral materials. Since Desert soils develop under so sparse a cover of widely spaced desert shrubs, they are very low in organic matter. This is the chief reason for their light color. Weathered rock material is widely exposed and imparts its color—usually red, brown, or yellow—to the surface soil. A large proportion of desert surfaces are not covered with mature soils. Instead, large surface areas are composed of bare rock, cobblestones, gravel, or shifting dune sand.

Some desert soils respond to irrigation. Within all desert regions are numerous intrazonal soils that have acquired excessive amounts of saline and alkaline materials. Salt flats of sodium chloride, borax, and other salts are widely distributed. Because of harmful salts, some Desert soils will not respond to irrigation. Some other Desert soils are highly productive under irrigation. Without irrigation all Desert soils have very low utility.

THE TUNDRA HAS LITTLE OR NO TRUE SOIL

In the strictest sense, the tundra of the Arctic fringe can scarcely be said to possess soil. Soil profiles, if we may speak of the surface accumulation as soil, show evidence of excessive rather than deficient moisture, even where precipitation is extremely light. During the long, cold winter the ground freezes solid from the surface downward, and all soil-forming processes are at rest. During the short summer when the surface layer is thawed, permafrost (permanently frozen ground) impedes downward drainage. Thus during the short, cool summer when soil-forming processes might have a chance to work, severe waterlogging often persists.

Tundra soils resemble feebly developed podzols. Better-drained tundra areas have soils resembling Podzols, with a brown, peaty layer covered by grayish horizons. But even these Podzol-like soils show very feebly developed profile features. Most of the tundra surface is made up of bog and hummocky marshland. Extremely harsh climatic conditions preclude agriculture and restrict natural vegetation to such species as mosses, lichens, and ferns. Man's use of Tundra soils probably will continue to be restricted to a limited pasturing of reindeer, sheep, and other similar domesticated animals and to a hunting ground for wild arctic animals.

AZONAL ALLUVIAL SOILS ARE WIDESPREAD AND HIGHLY VARIED

Most zonal soils have alluvial soils associated with them. In many, if not most, instances the alluvial soils are the most prized lands.

Few generalizations can be made about alluvial soils. Their materials, however, represent all varieties of color, texture, and plant-nutrient content. Relatively poor drainage is not a universal characteristic, since floodplains in arid regions often are too dry for crops most of the year. In general, alluvial soils are not highly productive. Some are too poorly drained; some flood during the cropping season; some are located in areas with a climate too cold or a growing season too short for agriculture; and some are composed of nearly sterile sand and gravel.

Organic matter usually collects rapidly. Alluvium may collect rapidly or slowly on a river flood plain. The loosely deposited materials usually are so thoroughly weathered and mixed that vegetation common to the area quickly establishes itself. If flooding is infrequent, as on second bottoms and terraces, organic matter accumulates, forming a surface layer darker in color and different in other ways from the layer underneath. In this way horizons have their beginning and youthful soils develop. Alluvium remains in undeveloped form as long as flood waters inundate it periodically. When flooding no longer recurs, for one reason or another, azonal soils develop rather rapidly into zonal soils, as they adjust themselves to the climate and vegetation characteristic of the area.

Shallow-rooted crops fit alluvial soils best. Alluvial soils, because of their relatively high water table, often are better adapted to shallow-rooted crops such as grains and grasses than to deep-rooted plants, especially those with a tap-root such as cotton. Much of the rice crop of the Orient grows on river alluvium. In many mountain areas the arable land is limited almost entirely to the coveted ribbon-like flood-plains and terraces of mountain streams.

Although alluvial soils are not maturely developed, and are not true soils in the sense of having evolved under the influence of soil-forming processes, they comprise a highly valuable natural resource. For countless generations they have provided nourishment for a large proportion of the world's population.

MOUNTAIN SOILS VARY WITH HEIGHT AND EXPOSURE

Soils of mountain terrain embrace a wide range of phases, types, series, families, and groups. There is no such thing as a

Fig. 5–11. Alpine meadow soil in the Medicine Bow National Forest near Laramie, Wyoming. Soils are thin and stony, grass is sparse, and grazing is limited to a short summer season in most mountain areas. Note the snow-capped mountains in the distance.

COURTESY OF SOIL CONSERVATION SERVICE, U.S.D.A.

"mountain soil group" or a "mountain soil class." In fact, individual soils of mountain areas are quite like those of other kinds of terrain. Many mountain soil types, series, and groups duplicate those found elsewhere, and are given the same names. Because of their elevations, mountains have lower temperatures than their adjacent lowlands. Thus, mountain soils resemble those of higher latitudes more than they resemble soils of their neighboring lowlands. For instance, in the higher reaches of the Appalachians, Podzols similar to those of southern Canada are found.

Nevertheless, mountain soils might be considered separately because they are somewhat misplaced, latitudinally, when compared with lowland soils. Mountain soils are characteristically immature, shallow, and stony, and they are disrupted frequently by protruding exposures of bedrock. But they have little else in common with each other. Individual soils reflect the local rock from which their parent material is derived. Local climates have their effects on soils. Soils vary from low to high altitude, from north slope to south slope, and from rainy slope to dry slope. Soils also vary as vegetation varies.

Basins, coves, and valleys provide choice mountain soils. Some mountain soils are more agriculturally productive than others. Basins, coves, and stream valleys usually provide considerable land with deeper soils and gentler slopes than found elsewhere in mountains. Calcareous rocks frequently provide parent material for basin and cove soils and thus enhance their value for crops. Subsistence agriculture, involving small-scale cultivation of steep slopes and narrow floodplains, is characteristic of many mountain areas. This type, however, sometimes gives over to commercial truck and dairy farming in larger mountain valleys and basins under favorable soil conditions.

Good trees and grasses grow in mountain soils. Pasture and tree crops, particularly apple orchards, frequently utilize intermediate slopes, leaving the steeper, higher slopes to permanent forests. Much of the mountain land of the United States belongs to the federal and state governments and is publically managed for forest, grazing, and recreational use. In Eurasia and Africa, alpine summer pastures have, since antiquity, played important roles in nomadic herding.

Word List for Study

distribution	dominance
topography	loam
humid lands	precipitation
tundra	evergreens
perennials	dormant
deciduous	subarctic
conifers	weathering
organic material	volcanic ash
oxides	diversification
Chernozem	Rendzina
Grumusol	flocculation
oxidation	soil profile
permafrost	sterile sand
subsistence agriculture	floodplain
Eurasia	

Questions for Consideration

1. Explain how topography may set the soil pattern of a particular area.
2. Why must we include climate and vegetation along with topography if we are to understand broad regions of soil patterns?
3. Which soils are formed under influences other than climate?
4. Why is the Hagerstown loam considered to be a highly productive soil?
5. What type of vegetation is usually found in the humid lands of the world?
6. How can you account for the fact that trees in the tropics do not shed their leaves regularly?
7. Explain why tropical soils are considered to be poor for agriculture.
8. Find out why soils of the middle latitude forests contain few insects.
9. Can you account for intense leaching in humid lands? Explain.
10. What kind of soil is usually found in cold, mountainous, forest areas?
11. Describe farming practices and types of fertilization and other treatments necessary for soils with strong acidity. Write a chemical equation to illustrate your point. (Review Chapter 1.)
12. List the crops which may be grown on Podzols.
13. Why is it possible for conifers to grow rather abundantly on Podzol soils?
14. Explain why the Gray-Brown and Brown podzolic soils from southern New England to Maryland are dark rather than characteristically gray like the true Podzol.

15. Describe the Latosols of the tropics. How are they developed? (Review Chapter 3.)
16. Why are the Latosols considered to be poor soils for agriculture?
17. Discuss the characteristics of the Cotton Belt soils of the South. Why does cotton grow well on these soils?
18. What kind of forest vegetation is prevalent in the Red-Yellow Podzols of the South?
19. How is moisture related to the Calcification process? Can you write an equation?
20. Why are Brunizem soils dark in color? What are the most abundant crops?
21. Contrast the Brunizem soils with the Rendzinas.
22. Describe the Chernozems. Why are they considered to be good agricultural soils?
23. What kinds of agriculture are practiced in Chernozemic-desertic areas? What crops are grown?
24. Why are Desert soils light in color?
25. Describe the Tundra soils. Are they useful for agriculture? What vegetation is usually found?
26. How are alluvial soils formed? What crops are usually grown on the alluvial fans?
27. What kind of farming is found in mountain soils?

CHAPTER 6

Uses and Care of Soils

HOW SOIL IS USED DETERMINES THE CARE IT NEEDS

Soils, like the lands with which they are associated, are used in a great variety of ways. In a broad sense, their uses encompass all the various ways in which land serves to fulfill man's needs and wants. Land economists recognize seven general classes of land use: agriculture, forests, recreation, mineral resources, water resources, transportation, and urbanization.

One sees clearly that soil as a resource is much more closely associated with the first two land uses, agriculture and forests, than with the other five. We cannot dismiss these others, however, since often they affect agriculture and forestry. For example, it is choice valley soils that are usually flooded (and so forced out of agricultural use) when a reservoir is developed to meet the need for water resources. Urbanization, transportation, and, to a less extent, mining and recreation often encroach upon agriculture and forests, depriving them of choice soils. On the other hand, subordinate land uses may contribute to an expansion of agricultural and forest acreages, as when agriculture is expanded under irrigation as water resources are utilized.

Agriculture is the leading soil use. For the world as a whole, agriculture, including pasturing, is by far the most significant class of soil use, measured in terms of area and population involved. More than one-half of the world's population of more than three billion are engaged in agriculture. A century ago the proportion was much higher. Less than one-tenth of the population of the United States lives on farms today, as compared with one-fourth in 1930. About three-fifths of the total land area of its forty-eight conterminous states is in farms, and less than one acre in six of this total land area is in crops; most of the remaining farmland, and much of the non-farmland, is in pasture and woodlands. In these forty-eight states, the proportion of all land in crops varies widely from one state to another. With 55.6% of its area in crops, Iowa has the largest proportion of cropland; with one-half of 1%, Nevada has the lowest proportion. North Carolina, which is about an average state in this respect, has 14% of all its land in crops. However, year by year, the proportion of cropland to non-cropland is decreasing in most parts of the country.

Most good soils are used for crops. There has always been great variation in the quality of soils available for agriculture. Many soils are well suited to a wide variety of uses. Others are suited to some general uses but not to more exacting ones. Some soils are not suitable for any agricultural use.

In long-settled areas much was learned about the qualities of soils by experimenting with different crops. Today the soils themselves are studied. What is learned through soil science greatly reduces the amount of chance involved in determining soil qualities compared to what was involved in old trial-and-error experimenting. In most instances, the better soils are used for crops, and the less suitable ones are kept for pastures or timberlands.

Good soils are used first. With the growth and spread of population, the better soils of new farms usually were cleared and cultivated first. Then, as more cropland was needed, less desirable lands with poor soils, steep and stony slopes, and poor drainage gradually were brought under cultivation. The drainage of wet lands and the irrigation of dry lands are old but costly practices in many areas. Such lands, however, are usually neglected longest and are used sparingly wherever good soils are plentiful. Much of the land that is undesirable for crops serves for pasture, woodland, wildlife, and recreation.

Little or no soil as such is needed for transportation, building sites, recreation, mining, and water resources. Land without true soil may serve these uses adequately.

Cropland requires greatest soil care. Properties that determine the agricultural quality of soils include: (1) ability to produce high crop yields under good management and careful handling; (2) the ease with which they can be used profitably; (3) the amount and kind of care they require. Good soils respond well to proper management, which involves correct cropping practices, use of fertilizer, and effective protection against damage. The increased yields on good soils that result from improved methods far exceed the extra cost of production in value, and thereby justify the added expense. Soils that are poor in some of their characteristics may fail to respond well to expensive handling methods and even worsen if neglected and abused. Thus without good care all cropland deteriorates with continuous use. The loss that results from improper care of good soils is greater than from improper care of poor soils, since the former are more valuable. Nevertheless, it is highly important to give the best possible care to all soils, and particularly to the best soils. This is one of the first essentials of good farming methods.

MOST CROPS ARE GROWN IN ROWS

Most of the common American crops and many crops of other lands are grown in rows. It does not follow, however, that all cropland is row tilled. A number of crops occupying vast acreages in America as well as in the other continents are sown in broadcast form or planted in closely spaced drills. The small grains, most hay crops, and rotation pasture largely comprise this group of crops that are not row tilled.

The more common American row crops are corn, cotton, tobacco, soybeans, peanuts, most truck crops, small fruits, and berries. Perennial orchard fruits and nut trees, grapes, and perennials for transplanting usually are grown in rows. Practically all row crops are intertilled (cultivated during the growth period). Occasionally, widely spaced rows of perennial orchard fruits, nuts, berries, and grapes are interplanted with "catch crops" of various kinds, either row planted or drilled. This is not a common practice, however, in American commercial orchard and vineyard culture.

Row crop methods vary little. General methods of row-crop tillage vary little from one crop to another. The chief variations are in frequency and depth of cultivation and in width of rows. Soon after seeds germinate and plants break through the surface of the soil, tillage usually is begun. It may start with harrowing or shallow cultivation with small plows. Where there are heavy rains, crusts may form, which in some soils may prevent or greatly hinder the young plants from coming through. Early cultivation is necessary to break the crust and to induce the circulation of air about the roots of the young plants. Also, early cultivation helps to control weed growth.

In areas of light rainfall, frequent cultivation helps to conserve moisture by keeping down weed growth and by checking evaporation. Cultivation with hand tools, of which the hoe is the most important in America, often is necessary. By hand cultivation, weeds beyond the reach of harrow or plow are removed, and plants are thinned to a proper "stand." With some row crops cultivation may continue at frequent intervals through most of the growth period until the crop approaches maturity.

Row cropping induces soil erosion. The cultivation associated with row cropping keeps practically the entire surface loose to a depth of several inches. This looseness makes the surface soil highly vulnerable to erosion by torrential summer rains. Most row crops do not develop a thick, wide-spreading mat of roots. The lack of this protection makes the entire soil surface even more subject to severe erosion unless careful precautions are taken. Since row crops require much more protection against erosion than do the "mat crops" with closely spaced plants, much of the work of soil conservation is concerned with the protection of row crops against soil erosion.

CLOSELY SET PLANTS WITH INTERLOCKING ROOTS FORM MAT CROPS

Mat crops consist of closely spaced plants with interlocking roots. These plants have spreading lateral roots that fill the top soil to depths of several inches. Plants with tap roots have few lateral roots to mat the top soil. As stated earlier, all small grains are classed as mat crops. Even corn, if grown in thick enough stands, forms a root mat. Pasture grasses and most hays are mat crops. Since mat crops are

Fig. 6–1. Harvesting wheat on Ulysses silt loam in Beaver County, Oklahoma. Small grains and other mat crops offer excellent protection against water and wind erosion.

grown in thick stands, matting the soil and covering the ground, they also are called "cover crops."

Mat crops provide protection against erosion. Mat crops protect the soil against erosion in two ways. Since their plants practically cover the ground surface, they serve to break the force of falling raindrops, thus reducing the pounding and splashing effects that are highly detrimental to bare soil. The more important role of mat crops in reducing erosion, however, is that of binding the soil together and holding the surface soil in place as runoff water flows down sloping surfaces during heavy rains. By measuring the amount of soil loss under known conditions of rainfall, slope, and soil character, it is possible to compare different kinds of soil cover as protection against erosion. It has been demonstrated that soils well covered with mat crops suffer the least erosion of any farming soils.

"SOD FARMING" SUBSTITUTES CHEMICALS FOR TILLAGE

Experimental work with chemicals in efforts to control certain plant growth has been carried on for many years. "Weed killers" of various kinds have been developed and are in wide use on lawns, railroad and highway rights-of-way, and other more restricted areas. The chemical "defoliators," especially those used on cotton just before the harvest, are another illustration of "chemical farming" that has advanced well beyond the experimental stage.

Among the most recent advancements in agricultural chemistry is experimental development of chemicals that are selective with respect to specific plants; that is, certain plants will react to such a chemical in a desired manner while other plants may not be affected by it. Such chemicals are called *herbicides*. Instead of killing plants outright, some herbicides are designed to arrest the growth of certain plants and hold them in a dormant stage for a desired period of time. The advantages of such chemical plant controls are numerous. For example, a sod of hay or forage crop may be held dormant three or four months, say from midspring until late summer, thus allowing time for a newly planted crop such as corn to mature. The sod crop may then burst into rapid growth with the summer rains and produce a full crop of hay or pasture after the corn is harvested. Experimental work is progressing rapidly, but it has not yet reached the stage of perfection hoped for in the near future. Also, costs are too high to make private use feasible on a large scale. Definite progress has been made, however, and widespread use of herbicides in the near future is predicted.

Sod farming offers several advantages. Sod farming is not yet practical except under highly favorable conditions. In most instances where it is practiced by private individuals some degree of public subsidy is available. It should offer several advantages once the herbicides are perfected and their cost lowered to the extent that this method becomes competitive with other farming systems.

The greatest advantage of sod farming over other methods is, perhaps, its protection against soil erosion. Slopes too steep for row cropping, even by contouring or terracing, lose very little soil under the practice of sod farming. This results from the fact that the ground surface is covered with a root mat at all times and there is very little tillage to stir the soil. Sod farming also makes feasible the tilling of a great deal of land that is not suitable for cropping by other methods. Newly cleared land can be sod farmed by planting among stumps and deadened trees. Land too wet for frequent tillage may be sod farmed in shallow-rooted row crops such as corn. Rough, stony land may prove better adapted to sod farming than to other methods.

SPECIAL CROPPING METHODS FOR ARID LANDS: IRRIGATION AND DRY FARMING

Practically all methods of crop raising including row cropping, mat cropping, sod farming, truck farming, orcharding, and grape culture require irrigation in arid lands.

Fig. 6–2. Pre-planting irrigation on Clovis fine sandy loam in Roosevelt County, New Mexico. For efficient use of irrigation water, some soils require leveling with large machines. Many soils of arid regions are highly productive when irrigated.

COURTESY OF SOIL CONSERVATION SERVICE, U.S.D.A.

There are several different methods of applying water for irrigation. The most common practice is furrow irrigation. This method is restricted to row-crop farming, in which water is released in furrows between the rows of plants. Flooding is a common practice where mat crops dominate. The levelness of the land is highly important to successful irrigation by flooding. The rice fields of the Texas-Louisiana coastal prairie are irrigated this way. The sprinkler method, in which water under pressure is discharged from overhead sprinkler pipes, is effective. It proves too expensive to be practical except where the farming is on a very intensive scale. Subirrigation consists of applying water through a system of porous pipes buried beneath the surface. It is the least practiced of the common methods of irrigation.

Dry farming alternates cropping and fallowing. Dry farming is more than merely farming in arid regions. It is a system adapted to the particular natural environment of each arid region in which it is practiced. Dry farming consists of crop raising and fallowing in some definite alternating sequence. A common practice is raising a crop in alternating years. In some instances crops are grown in two consecutive years and the land is allowed to lie fallow in the third year. Regardless of the cycle used, the objective is to conserve moisture during the fallow year for crop use the following year.

Dry farming methods are by far the most common for wheat raising in America. On the Columbia Plateau of the State of Washington, dry farming attains its highest degree of development. Under favorable conditions, wheat is produced there with as little as 8 inches of annual rainfall.

CROP ROTATION CONSERVES SOIL AND IMPROVES FARMING

The systematic alternating of crops from field to field is known as crop rotation. A good rotation system involves a great deal more than a haphazard switching of crops. It consists of adjusting the crop arrangement to the physical nature of the land and, at the same time, maintaining a balanced economic farming program. On deep, fertile soils with a gentle slope, a two- or three-year rotation may be satisfactory. On steep land with thin, stony soils, a rotation period of four to eight years may be most desirable. Rotation implies the growing of more than one crop on a farm. In other words, rotation and diversification go hand in hand. The combination of the two practices has meant a great deal in the advancement of American agriculture, particularly in the South. They help maintain the soil at highest possible levels of productivity; they contribute to a balanced farm economy, and they help distribute farm labor and income throughout the year.

Since no two crops make identical demands on the soil, one crop may require excessive amounts of a given nutrient which another crop may be able to supply. For example, legumes return nitrogen to the soil, but most other crops require more nitrogen than the soil can normally supply. Thus a rotation program involving clovers, cowpeas, and other legumes combined with cotton, corn, small grains, and other heavy users of nitrogen may meet most nitrogen requirements.

Rotation helps balance the farm program. Crop rotation and associated diversification help improve the economic status of farming in many ways. Besides maintaining the soil in a good state of productivity, rotation and diversification help reduce the risks of economic losses from low yields, low prices, or a flood disaster. Also, the income and the labor requirements are spread through the year instead of being seasonal in nature, as in the case of a single crop, such as wheat. Some phases of a well-ordered crop rotation program are apt to succeed, even if others fail. It is not necessary for all parts

Fig. 6–3. Ranch scene in Roosevelt County, New Mexico. The soil is Amarillo loamy fine sand belonging to the Reddish Chestnut group. Grazing is less demanding of soils than is crop agriculture, yet soil quality is important in determining the value of the pasture.

COURTESY OF SOIL CONSERVATION SERVICE, U.S.D.A.

of the program to show a financial profit; it is the overall result that counts. A soil-improving crop, for example, may not be economically profitable at the time it is grown, but its benefits may be clearly realized in future years.

PASTURES EXACT LESS OF SOILS THAN CROPS DO

Much that was said in the foregoing pages about crops applies equally well to the use of pastures. All land employed in the rotation of pastures is regarded as cropland and must be handled as such. Much permanent pasture has come into use by first treating the soil with practices designed to improve it, such as clearing timber from it, fertilizing, liming, and sometimes draining it. Vast acreages of permanent pasture, however, consist of lands that are not well suited to crop culture. Rough relief, thin, stony soil, and light forage cover describe much pasture land, particularly in the rainier parts of America.

Soil is less significant in pasture than in crop uses. Even in those pasture lands that are not suited to crop use, soil often is a highly important factor in pasture quality, good soils giving rise to good pastures, poor soils producing inferior pastures. In the study of the Calcification process of soil formation (p. 52) it was shown that grasses are heavy users of bases and that the best grassland soils are well supplied with lime and other bases. Thus it follows that many choice natural pastures are found in areas with soils of limestone origin. The Kentucky Blue Grass Basin, the Nashville Basin, the Alabama Black Belt, and the Texas Black Prairies are examples of excellent natural pastures. Good pasture lands, however, are usually also good croplands, unless their roughness, stoniness, or shallowness make them unprofitable to farm. Therefore, the choice lands are planted in crops; and, usually, only the less desirable areas are utilized for pastures.

For the most part, the vast grasslands of the middle latitudes are located in areas of light rainfall. Tropical grasslands have a heavier rainfall that is distinctly seasonal in distribution. In either case, grasses prevail over trees, because the latter require a moisture supply throughout the year. Although the rainier grasslands have climates suited to crops, in semi-arid and arid lands, crop agriculture usually gives way to grazing. Thus in many areas, regardless of soil qualities, grazing is the only type of agriculture that is feasible.

SOIL CARE: CONSERVING AND IMPROVING

The foregoing pages were concerned mainly with the many ways in which soils are used in modern agriculture. In a general sense, most agricultural soils in use today have been in constant use for long periods of time; most of them must continue indefinitely in similar uses. The Chinese have farmed some of their land for forty centuries and may continue farming it in a similar manner into the indefinite future. Land use is a continuous operation. Farmers normally do not put their land aside while they are restoring their soils; they repair and improve them while the land is in use. The ways that soils are used, the length of time they remain productive, and the harvests they yield depend to a marked degree upon the care they receive.

Time makes little change in virgin soils. But when soils are used for crops or pasture, the balance that nature has given them is upset in various ways and to varying degrees. Changes in the nature of soils cannot be avoided as they are put to divers uses. These changes may result in improvement in productivity. Frequently, however, soil use results in soil damage and decreased yields. Thus careful treatment of soils in ways that will keep them productive through continuous use is the aim of every good agricultural program.

Prevent damage and repairs are unnecessary. It was shown in Chapter 1 that soils are like living things in many respects. Although soils cannot reproduce themselves, they possess the unusual ability to sustain the normal life of plants and animals without deteriorating. Under natural conditions soils never wear out; under careful and proper use they may continue indefinitely to produce good yields without wasting away or losing their productive powers. Soils even "make minor repairs," healing their own wounds under constant careful use. It is improper use and lack of care that harm soils, often damaging them beyond repair. Thus good care is of vital importance in prolonging the useful life of soils; if damage can be prevented, remedial measures are unnecessary and soils continue producing for generations.

Erosion control is a first measure. Erosion perhaps is the most damaging thing that can happen to agricultural soils. Severe gullying is destructive in two ways. Gullying often destroys the entire soil, stripping away the whole profile down to bedrock or underlying parent material. Again, gullying may so completely cut a field to pieces that the remaining uneroded soil cannot be cultivated economically. Thus a few deep gullies may ruin an entire field or farm.

Sheet Erosion is the gradual removal of a thin surface film of soil. It may proceed so slowly that it is hardly noticed. (Muddy water draining out of a field proves it.) Yet if no more than an eighth of an inch of soil is lost annually through sheet erosion, the entire A horizon may be lost within one generation of farmers.

Wind Erosion, much less widespread and damaging than water erosion, is serious in most arid lands and in rainier areas where wind velocities are high. Soil damage from wind erosion results both from the removal of surface soil materials and from the deposits of them left elsewhere. The fury of duststorms and sandstorms is familiar to most people. But the great damage such storms do to soil, which is lasting and costly, may escape general notice.

All soil erosion is harmful and expensive. It removes or destroys soil material that cannot be replaced. It upsets the balance that nature has worked through pedologic time to produce. Soil erosion may also be harmful to other lands down slope or down wind where eroded materials are deposited. Checking erosion and repairing the damage already done is a must and a first measure in scientific agriculture.

Repairing damage follows checking erosion; they may proceed together. Checking erosion is the first essential of soil conservation. Little is to be gained by restoring badly eroded soils unless erosion is first arrested and further erosion prevented or greatly reduced. Both processes, however, may go on simultaneously. For example, old gullies may be filled as terracing and contour plowing proceed. Legumes and other soil-improving crops may prove equally effective in controlling erosion. Whether soil damage is advanced or just beginning, checking it is a necessity.

Preventing other soil damages requires constant vigilance. Other soil damages may be as serious and as costly as erosion, though none is as widespread. Improper use may result in the deterioration of soil structure; several things may contribute to this deterioration, including: plowing when soil is too wet; failure to return organic matter; unwise use or lack of lime; neglecting to rotate crops. An even more serious and widespread kind of soil damage is the loss of essential plant nutrients. This may result from continuous growing of the same crop and from failing to fertilize it properly. Still another serious cause of depletion in most soils is the loss of organic matter. Clean harvesting can be more costly in the long run than plowing under stands of grain, hay, or pasture.

Most kinds of soil damage are related to each other in many ways. For example, organic deficiency is definitely a factor in soil erosion and in structural breakdown as well as in the deficiency of plant nutrients. Consequently, understanding these intricate relationships and properly evaluating them in a balanced farming program make modern agriculture truly a scientific undertaking.

Soil care is more than soil protection. Most soil care is directed at correcting damage already done and preventing deterioration in productive power. It must be remembered, however, that many soils are capable of improvement far beyond their virgin qualities. By such practices as draining, liming, leveling, and desalting, many almost worthless soils have been

Fig. 6–4. Soil care is vital to successful farming. This field of Blount silt loam in Madison County, Indiana was plowed too wet.

COURTESY OF SOIL CONSERVATION SERVICE, U.S.D.A.

made productive. Widescale use of commercial fertilizers has contributed most to increased crop yields; trends show highly successful results from their use on some crops, including corn, cotton, and tobacco. The best results from fertilizers, however, are associated with improvements in farming methods and management.

SOIL CARE CAN BE A MANAGEMENT PROBLEM

Some of the steps in designing a farm plan are directly concerned with soils, while others involving such matters as labor and marketing relate to soils only indirectly. Every good farm plan must take into careful consideration the characteristics of both the land and the soils. The entire program must be based on these natural features, since they are strongly related to the desirable kinds of uses. The soils on most farms vary considerably with respect to their feasible uses, or "capabilities."

A good farm program may well begin with making a map showing the different classes of land. Such a map is not easily drawn. It requires the services of a trained soil conservationist who understands both soil science and agriculture. These "pub-

Fig. 6–5. A North Dakota farm scene showing contour strip cropping and pasturing. Soils of the area belong to the Chestnut group. Crops shown are alfalfa, wheat, and corn. This system facilitates both erosion control and crop rotation.

COURTESY OF SOIL CONSERVATION SERVICE, U.S.D.A.

lic servants" are available in practically every farming community. Farmers should not hesitate to use their expert services.

The illustration within Chapter 4, figure 4–5, that shows classes of land capability is broad enough to cover all possible land uses. No single farm is apt to have all eight capability classes represented; however, almost every farm has several classes. This illustration, then, represents a hypothetical situation to show the significance of each class as defined in the legend. The farm plan map, figure 4–6, illustrates a plan for a particular farm, based on information concerning its use capability as it is shown in the land capability map, figure 4–5. From the illustrations and the legend describing the terms, one sees how complex a farm plan based on land capabilities may become. On the other hand, he sees the advantages of such a plan, if expert assistance is used in drawing the plan and fitting the farm program to it.

Following a farm plan requires cooperation and advice. A good plan is only the beginning. Following the plan calls for efficient management as well as careful, industrious labor. Cooperation and sound advice are most valuable if used wisely. Soil conservationists, county agricultural agents, agricultural experiment stations, and other "public servants" generously provide assistance to the farmers who will use it.

GOOD SOILS RESPOND WELL TO A VARIETY OF TREATMENTS

Through experimentation much has been learned about the responses of individual soils to many kinds of treatment. The extent to which a soil responds favorably to "recommended" methods of handling and treatment is an important criterion in judging its value as an agricultural soil.

Soils with desirable qualities such as good texture, structure, depth, tilth, gentle slope, and an abundant supply of plant nutrients are classed as good or excellent agricultural soils. As a general rule, soils possessing these qualities also make excellent response to proper management, tillage, ferti-

Fig. 6–6. Good soils bring high returns on expensive treatment. This North Carolina field of tobacco is on Norfolk loamy sand. With heavy fertilization and other good practices, a good tobacco crop may gross as much as $1,800 per acre with a favorable market. Notice the sweet potatoes between the tobacco rows. Every fifth row is planted in potatoes; and, after the tobacco is harvested, hogs feed on the root crop.

PHOTO BY KEITH D. HOLMES.

lization, and similar practices. One way of measuring a soil's response to good treatment is to compare yields under controlled levels of treatment. Soil scientists have rated the soils of some areas with respect to yields under "no" treatment, "usual" treatment, and "recommended" treatment. In most instances soils that are good to begin with make the more profitable responses to good treatment.

Good soils bring high returns on expensive treatments. The more expensive the soil treatment, the more urgent the need for a profitable response. Expensive irrigation and fertilization, for example, increase profitability more when applied to the best soils. The best soils are most often used for the most profitable crops, especially under an acreage control system. Thus there is a double reason for concentrating efforts and expenditures on the choice soils.

Tobacco farming illustrates this principle. An excellent tobacco soil, such as the Marlboro sandy loam or the Norfolk fine sandy loam, under "recommended" practices yields a gross income in excess of $1800 per acre. The average gross return under "usual" practices drops below $1200 per acre. With no treatment of any kind the return is far less and money is lost when the crop is grown. Although production costs are highest under the "recommended" practices, the increase in gross returns brings greater profit.

The best soils usually require the least protection. Another measure of soil quality is the extent to which it requires protective care against deterioration. Since erosion is the greatest destroyer of soil, the farm planner may well start with the erodability of soils. As was pointed out on p. 136, a farm plan map logically begins with land capability classification (figures 4–6 and 4–7). In some respects a capability classification map is at the same time an erodability map, since each land capability class requires a different degree of protection against erosion (and, to a limited degree, wetness). The best soils (capability classes I and II) require practically no protection against erosion. This is largely due to the very low degree of slope. Soil character also is important, since individual soils differ greatly in erodability and natural drainage, even with identical slope and rainfall.

Further examination of the land capability classification reveals the strong emphasis on soil erodability in the higher numbered categories. Classes VII and VIII are too erodable and too rough for agricultural use. Intervening classes III, IV, V, and VI are recommended for crops and pasture, but require varying degrees of protection against erosion.

It does not follow that all soils free of erosion problems are good agriculturally. For example, coarse, sandy soils with rapid interval drainage have no erosion or drainage problems, yet they usually make very poor soils for crops and pastures.

Many good soils have erosion problems. Although the best soils are least erodable, in general, there are many examples of good soils on relatively steep slopes that require careful protection against erosion. This is true in the better farming areas of the Appalachian Piedmont (e.g. Lancaster County, Pennsylvania), the Great Appalachian Valley (c.g. Shenandoah Valley), and the Kentucky Blue Grass.

SOIL CONSERVATION: A WORLDWIDE NEED

Soil Conservation is a momentous term, for it means conserving man's most important natural resource. Its three goals are to check or prevent soil damage, to repair damage already done, and to maintain as high a level of productivity as possible while the soils are in constant use. The need for soil conservation exists wherever farming is practiced. The application of soil conservation practices on a systematic, organized scale has not yet reached all areas that need it, but it is continuously spreading.

Many books and articles have been written about soil conservation. Courses in the subject are taught in high schools and colleges. Many people spend much of their time in conservation practice and the public is constantly reminded of the need for it. In this book, however, the limited space permits only a very brief treatment of soil conservation.

Erosion control is of major importance. The emphasis on erosion control is natural in a soil conservation program. Such an approach recognizes erosion as the most destructive force that affects soils. It also recognizes the fact that practically everything else done to improve soil is wasted effort if erosion continues unchecked. The destructive effects of soil erosion need no emphasis here.

The common practices of erosion control are familiar to most students. Several of these practices have been mentioned already in this book. The list includes terracing, contour plowing, cover cropping, strip cropping, and crop rotation. Each practice has its particular place in a soil conservation program.

Expert advice in applying the appropriate practice is always important. Soil conservationists, county agricultural agents, and other specialists stand ready to assist the farmer in matters concerning soil conservation.

Soil maintenance must accompany erosion control. The full value of erosion control can be realized only if the soils are kept in good condition. Once erosion is brought under control, soils may be used continuously with good results if the appropriate methods are followed. Maintaining soils at a high level of productivity, and even improving them beyond their former level, is possible through efficient fertilization, crop selection, tillage, crop rotation, and other methods of soil maintenance and improvement.

Soil maintenance and erosion control must go hand in hand. Neither accomplishes much alone. Wise planning and good management are essential. Expert assistance is always available to interested persons.

Soil conservation is a national problem. Soil conservation has become a national problem in America. The Federal Government began working seriously at it in the 1930's. The Soil Conservation Service, established in 1935, has become one of the United States Government's largest services. It operates to assist farmers in every part of the country in which agriculture is practiced. State and local organizations work with the federal agency. Together they carry their valuable services into all local farming communities that will accept them. A great deal of good has been accomplished by the Soil Conservation Service and the many cooperating units, and their programs are expanding greatly. However, the work of this agency and the others participating in its program will never be finished; it must continue indefinitely.

Everyone benefits from soil conservation; everyone should assume responsibility for it. Soil conservation is mainly the problem of the farmers, but all of us depend upon the soil as the source of most of our food and of many other products. As citizens we all share responsibility for its conservation. All of us can contribute something to soil conservation; some can contribute more than others. Out of student groups of today will come the leaders, and many of the followers, of tomorrow. There is good reason to be optimistic about the future of American agriculture, but we must be realistic. We must accept the proposition that soil conservation is essential if our agriculture is to advance and to meet the challenge of the future. Let every citizen, then, face his responsibility. Let each one of us be a "conservator" at heart.

Word List for Study

irrigation	diversification
soil deterioration	perennial
intertillage	vineyard culture
mat crops	lateral roots
sod farming	herbicide
defoliation	contour
terracing	dry farming
gullying	erodability
soil quality	

Questions for Consideration

1. What are the general classes of land use recognized by land economists?

2. Investigate and explain why many people are leaving the farms. How can food production continue to meet the needs of our expanding population?

3. What constitutes a good soil? Why are the poorer soils used for pasture and timber production?

4. Describe some good practices in soil management. How can cropland deteriorate?

5. Distinguish between row crops and crops sown in broadcast form.

6. What is the purpose of harrowing and shallow cultivation? How can frequent cultivation conserve moisture?

7. How does row cropping make soils vulnerable to erosion?

8. Define the term "mat crop." How is mat cropping related to soil conservation?

9. List several mat crops. Why is corn considered to be a mat crop?

10. Describe "sod farming." Why is its use limited?

11. How does "sod farming" provide protection against soil erosion?

12. Discuss the various methods used to irrigate crops.

13. What is the main purpose of crop rotation? Distinguish between rotation and diversification.

14. Why are the best pastures grown in areas with soils of limestone origin?

15. Illustrate ways in which improper soil use can produce severe damage to soils.

16. How can soil erosion be prevented? List several ways.

17. Investigate what may be done to prevent erosion in geographic areas where dust storms and sandstorms are prevalent.

18. What is meant by "land capabilities"? Why is farm planning so important?

19. Discuss the relationship of soil qualities to crop yields.

20. Why is soil conservation considered a worldwide need?

21. Why should each person, as a member of the world's society, be a "conservator"?

Chemistry of the Earth

THE EARTH IS NEVER STATIC

Meteors, meteorites, and meteoritic dust from outer space are constantly falling onto the surface of the earth. At the same time, atoms of gases such as hydrogen and helium are being released by the atmosphere of the earth, the gravitational pull of which is not great enough to hold all these atoms of gases near its surface. The incoming particles of meteors and meteoritic dust may be considered as gains, while the release of gases may be regarded as losses. However, these gains and losses are insignificant in comparison with the entire mass and atmosphere of the earth.

Indeed, constant changes are taking place in the earth's interior and in its crust, as well as in its atmosphere and in outer space. Some of these changes are observed by both scientists and laymen, but other changes are not noticed. For example, we have no direct evidence of any alterations in the earth's core, and our theories about such changes are purely speculative. But by studying the physics and chemistry of the crust, we can gather sufficient scientific data on which to base our working hypotheses.

Almost all scientists agree that the earth was neither created nor formed in the state in which it now exists, but that it has evolved as a result of slow chemical and physical reactions that have been gradually taking place within it and on its crust for more than three billion years. Geochemists are more concerned with the earth's surface than with other parts of it, for its surface is more readily accessible to direct observation.

By observing the different kinds of rocks, we can see that various activities in the earth's crust are interrelated. Magnetism, for example, is related to igneous rocks, sedimentation to sedimentary rocks, and metamorphism to metamorphic rocks. To understand the different properties of the earth's crust, we must begin with the laws of chemistry.

DEFINITION AND PROPERTIES OF MATTER

As a student looks about his classroom, all the objects that he sees are composed of what scientists term *matter*. The door he closes and the chalkboard he writes on are different types of *solid* matter. The water he drinks is *liquid* matter, and the air he breathes is *gaseous* matter. All three types of matter occupy space and have weight. Such characteristics as *mass* and *weight* are called the properties of matter. However, there are many kinds of matter, and each kind has its own particular properties.

If a student were asked to describe a rock, a piece of glass, or a metal cylinder, he would probably do so by giving an account of its *physical* properties: a rock may be coarse grained, fine grained, or have no grain at all; a piece of glass may be either rough or smooth; metals may be hard or soft, bright or dull, hot or cold. If the student identifies a substance by its texture, color, weight, shape or other apparent characteristics, he is identifying it by its physical properties. But all substances have *chemical* properties as well as physical ones. For example, if a strip of metallic zinc is immersed in hydrochloric acid, bubbles of gas are released by the reaction between the metal and the acid. If left in a sufficient quantity of hydrochloric acid for a given length of time, the zinc will be dissolved, and new compounds with new properties will be formed. Metallic zinc has certain obvious physical properties, such as its blue-white color and its hardness and brittleness at normal temperatures, but not until it reacts with the acid does the chemical property described above become apparent.

The physical properties of a substance. For our purposes here, we need to consider ways in which a quick, practical identification of a substance may be made from its physical properties, although no complete, accurate identification can be made without a chemical analysis of its composition.

All matter exists either as a *solid, liquid,* or *gas.* A solid has a form of its own, while a liquid takes the form of the part of the vessel it occupies. A gas expands to take the form of the entire containing vessel. Most substances may be found in each of the three states; for example, water, ice, and steam.

We may ask ourselves a number of questions about matter: What is its color, its odor, its taste? How hard is it? At what temperature does it boil? Does it conduct electricity? Does it conduct heat? The answers to these or other similar questions enable us to describe or identify the *physical properties* of a substance.

The chemical properties of a substance. The chemical properties of a substance are the qualities that determine whether or how it reacts with other substances. When iron comes in contact with oxygen, it rusts; when silver comes in contact with sulfur or sulfur compounds, it tarnishes. In both rusting and tarnishing, new substances with new properties have been formed as a result of the chemical reactions that have taken place. Thus the rusting of iron, the tarnishing of silver, the burning of oil, the decay of wood all take place because of certain chemical properties inherent in these substances.

THE COMPOSITION OF MATTER

A substance that cannot be broken down into a simpler substance by ordinary chemical means is called an *element.* In 1774, an Englishman, Joseph Priestley, heated a substance that he called mercury ash. As a result of heating, he obtained mercury and oxygen. No matter how much he tried, he could not obtain anything else because he had broken down the mercury ash into its simplest forms. A substance, then, that contains only one kind of matter is an *element,* for it is in its simplest form.

The story of the discovery of the elements stretches across several centuries of experimentation by chemists and physicists. Only twelve of the elements were classified as early as 1630. Only one of those twelve, iron, is among the twelve most abundant elements of the earth's crust.

Elements Known in 1630	*Most Abundant Elements of the Earth's Crust*
antimony	aluminum
bismuth	calcium
carbon	chlorine
copper	hydrogen
gold	iron
iron	magnesium
lead	oxygen
nickel	potassium
silver	silicon
sulfur	sodium
tin	phosphorus
zinc	titanium

An atom is the smallest particle of an element that can take part in a chemical change without being changed itself. It is so small that it cannot be seen with the most powerful microscope. Professor W. R. Whitney illustrates the minuteness of the atom by saying that one drop of water contains so many atoms that if each atom were as large as the drop itself, they would cover the whole earth with a foot of water.

Although atoms are too small to be seen, they do have weight. Atomic weights are very difficult to calculate. An atom of oxygen has been estimated by chemists to weigh .00000000000000000000000026 (2.6×10^{-23}) gram. This number is too small to work with conveniently, so chemists adopted a system of relative weights. Oxygen is the most active element since it unites with more elements than any other. Therefore, the weight of oxygen was used as a standard with which to compare the weight of other elements. Hydrogen is about 1/16 as heavy as oxygen. In order for hydrogen to have a weight of approximately one, oxygen was given the weight of 16. What this actually means is that oxygen is sixteen times heavier than hydrogen. When oxygen has a weight of 16, helium has a weight of approximately 4, because it is about 1/4 as heavy as oxygen. Carbon has a weight of approximately 12, uranium approximately 238, and so on. For the last several years physicists have used carbon rather than oxygen and hydrogen as a basis for determining weights. Since physicists and chemists did not agree entirely, two atomic scales have been used. In August 1961, the scale using the carbon isotope of mass 12 as a standard of comparison was approved by the International Union of Pure and Applied Chemistry and the International Union of Pure and Applied Physics. Now all scientists use this scale. The approximate weights are called *atomic weights* and are shown in the periodic table of the elements. A periodic table appears on page 292.

   The structure of the atom. Small as atoms are, within them are tiny subatomic particles called *electrons, protons,* and *neutrons*. An electron is a minute particle that has a negative (−) charge of electricity and a mass approximately 1/1836 that of a hydrogen atom. A proton is much heavier than an electron; it has a mass about 1800 times greater than an electron, and it carries a positive (+) charge. A neutron is an elementary particle having no electric charge (that is, it is electrically neutral) and a mass only slightly greater than that of a proton.

   What follows is a useful description of the atom. The nucleus of the atom is its dense, positively charged core; it contains one or more protons and, except in the case of hydrogen, one or more neutrons. Moving in circular or elliptical paths or orbits around the nucleus are electrons in sufficient number to make their total negative charge equal to the positive charge of the nucleus. These positive and negative charges counteract each other; consequently, the atom as a whole does not have an electrical charge; it is isoelectric, that is to say, it is electrically neutral.

The number of protons in the nucleus of the atom of an element determines the *atomic number* of the element. From the periodic table at the end of this chapter, it can be seen that the number of protons in the atoms of known elements

STRUCTURE OF THE SODIUM ATOM

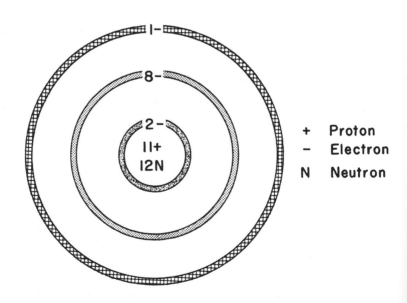

+   Proton
−   Electron
N   Neutron

THE SODIUM ISOTOPE

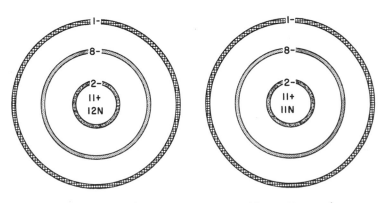

(Twelve Neutrons)   (Eleven Neutrons)

ranges from one in the hydrogen atom to one hundred three in the Laurencium atom. The number of protons in the nucleus, and therefore the atomic number, is equal usually to the number of electrons in orbit around it.

There are never more than two electrons in the first orbit nor more than eight in the outer orbit. For example, sodium has an atomic number of eleven because the nucleus of the sodium atom contains eleven protons. Two electrons are found in the first orbit, eight in the second, and the remaining one is on the outer orbit. It revolves in this orbit by itself. As we shall see later, this orbit determines the activity of the element.

Some of the sodium atoms have an atomic weight of 22 and some an atomic weight of 23. Since there are more of those with a weight of 23 than of 22, the atomic weight of the sodium atom is given as 22.997. *Isotope* is the term applied to one of two or more atoms of an element whose nucleus contains the same number of protons as the other atoms but a different number of neutrons. The isotopes of an element have the same atomic number but different atomic weights.

STRUCTURE OF THE CHLORINE ATOM

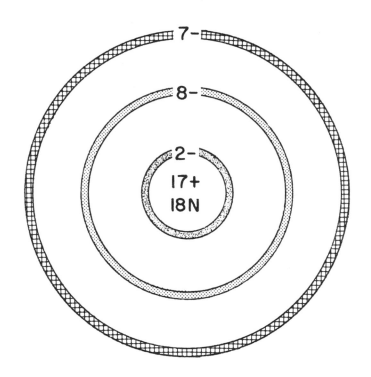

HOW ELEMENTS COMBINE

Elements are combined by a process of exchanging electrons. In the process the outer orbit of each atom is made complete when the atom either gains or loses electrons. The outer orbit of an atom is complete if there are eight electrons present. Some elements have atoms with eight electrons on the outer orbit. These do not combine with other elements. Argon, neon, krypton, xenon, and radon are examples; they are known as inert or rare gases.

The atoms of most of the elements do not have complete outer orbits. When the elements combine, the outer orbits become complete in one of two ways. If there are four or more electrons in its outer orbit, an atom will gain electrons to complete the orbit. If there are four or less electrons in its outer orbit, an atom will give up those electrons, thus leaving the next orbit, which contains eight electrons, as the outer orbit. For example, the sodium atom was described as having one electron on the outer orbit and eight on the next orbit. The chlorine atom has seven electrons on the outer orbit or shell. When sodium and chlorine (each in a free state) are placed together in a closed container, an exchange of electrons takes place and sodium chloride is formed. The single electrons from the outer orbits of sodium atoms are lost to the outer orbits of the chlorine atoms. The outer orbit of each chlorine atom is completed by gaining an electron. The next orbit of each sodium atom, with eight electrons, becomes its outer orbit and is complete.

REACTION BETWEEN SODIUM AND CHLORINE

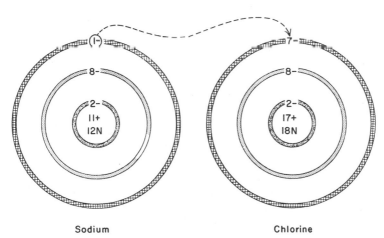

Sodium   Chlorine

The measure of the capacity of an atom of a given element to combine with the atoms of other elements to form a molecule is known as its *valence*. The term valence may also be used to indicate the number of electrons that an atom gains or loses in reacting with other atoms. The valence of an element may be positive or negative, depending upon whether its atom gives or receives electrons. Sodium chloride is an example. In its free state, the sodium atom has the same number of protons and electrons and is, therefore, electrically neutral; that is, it has no electrical charge. Refer now to our diagram on p. 151 and see what happens when an atom of sodium combines with an atom of chlorine. The sodium has one electron in its outer shell, which it gives away; thereafter, because it has one fewer electron than protons, it is left with a positive charge of one. Such positively charged sodium atoms are called sodium *ions*. Chlorine, on the other hand, has taken on the one electron from sodium and has become negatively charged. It is now a chlorine *ion*. When elements combine in this manner, they are known as *compounds*. Atoms of elements may exist by themselves, but ions are always part of compounds. Aluminum has a valence of three since it loses three electrons when it combines with another element. Oxygen exhibits a valence of minus two for it takes on two electrons. Therefore, it will take three oxygen atoms to join with two aluminum atoms to form Al_2O_3 (aluminum oxide). Can you diagram this transfer? What happens when calcium joins with chlorine?

The valence of metals and nonmetals. We have seen from the previous discussion that some elements have a positive valence and some have a negative valence. The *metals* lose electrons when they join with other elements. They cannot join with one another, for each metal must lose electrons. To unite chemically, one element must lose and another must gain, so we can readily see why metals must join with elements that gain electrons. It is possible, however, for some metals to have more than one valence. Iron, for example, may lose three electrons under a given set of conditions and exhibit a valence of plus 3 ($+3$). Under other conditions, iron may lose only two electrons and have a valence of plus 2. Some typical examples of metals are iron, aluminum, copper, calcium, sodium, magnesium, potassium, tin, lead, and zinc.

The *nonmetals* usually gain electrons when they unite chemically with the other elements and therefore exhibit a negative valence. Some elements may have both a positive and negative valence. Carbon is a typical example. When carbon unites with oxygen to form carbon dioxide (CO_2), it loses four electrons thus having a valence of plus four. When it joins with hydrogen to form CH_4 (methane), it gains four electrons, thus having a valence of minus four.

Notice the diagram of the carbon atom. Would you expect carbon to have both valences since it has an orbit one-half full of electrons? Some of the most common nonmetals are oxygen, nitrogen, chlorine, bromine, sulfur, and phosphorus.

Compounds and mixtures. As was shown above, elements may exist in the free state or they may be united chemically into a different substance. When two or more elements unite in this way, they are called *chemical compounds*. Examples of elements in the free state are silver, sodium, gold, and sulfur.

Not all substances unite chemically; they may occur together in the same place in nature. Sulfur and sand may be mixed. That is, the individual grains of sand and the crystals of sulfur may be mingled in such a way that they appear to be one. Actually, the particles of each may be identified and separated by mechanical means. In this case, they represent a *mixture*. In a compound, the elements are so closely associated with each other that the individual identity of each element is lost. This is not true, however, with a mixture, for the individual elements may be recognized by their physical qualities and can be physically separated.

Solutions. Many substances have the peculiar physical property of forming *solutions*. In most cases, solutions may be classified as mixtures:

(1) We may dissolve a teaspoon of salt in a cup of water. These two substances are miscible because they can be mixed together to form a solution. Their identities have been lost, yet no reaction has taken place between the salt and the water. The properties of each constituent have not been altered just because they have become a part of the solution. How may the salt and water be separated?

(2) When carbon dioxide is dissolved in water as, for example, in the process of making carbonated drinks, the carbon dioxide enters into solution to form carbonic acid, a weak, unstable acid, which makes the beverages effervescent. Since carbon dioxide can be kept in solution only at very low temperatures or under some type of pressure, caps are applied to such beverages during the bottling process. What happens when the cap is removed?

(3) A considerable quantity of sugar is mixed with very hot water. Almost at once, the sugar seems to disappear and becomes a part of the solution. Will the sugar remain in the solution when the temperature of the water is lowered?

Mixtures that are not solutions. A solution may be defined as a mixture but from the illustrations above we can see that it is a special kind of mixture. A solution of salt and water or sugar and water is very different from a mixture of sand and water. By *filtration* sand may be separated from the water. This is not true in the case of sugar or salt and water, for salt or sugar particles pass through the filter along with the water. (An exception is finely divided clay particles, which may pass through filter paper when mixed with water because the particles are so fine; nevertheless, they are not in solution.)

In order for a true mixture to exist, the components must be uniformly distributed throughout the mixture. A solution, therefore, is a homogeneous mixture throughout of two or more components.

Classes of Solutions

1. Solids in gases
2. Solids in liquids
3. Solids in solids
4. Liquids in liquids
5. Liquids in solids
6. Liquids in gases
7. Gases in gases
8. Gases in liquids
9. Gases in solids

How many examples can you give?

Solutions play an important role in the formation of rocks and mineral deposits. Water carries dissolved particles in solution until it reaches lakes, basins, or oceans. There the water evaporates and these particles are left on the bottoms. During the course of time, these deposits may become rock.

Classes of chemical compounds. For our purpose, it is not necessary to classify all the compounds in chemistry. In inorganic chemistry alone there are more than 30,000 separate compounds, and each compound exhibits its own physical and chemical properties. We can see how impossible it is to learn about the individual compounds. In general, we are most concerned with certain types of compounds which are encountered in the study of rocks, minerals, and soils: oxides, acids, bases, and salts.

Oxides. Oxygen reacts with various substances to form *oxides*. At ordinary temperatures, this process is slow. Iron, for example, will oxidize very slowly with oxygen to form a corrosive substance. This is known as slow oxidation. Can you name the oxide which is formed? What is its formula?

When the temperature is increased, the process of oxidation is speeded up. Ordinarily wood will not oxidize at a low temperature, but when it is heated the wood will burst into flames. This is rapid oxidation and is often called *combustion* or burning.

Billions of years ago, the earth was in a molten state. Since oxygen is very active at high temperatures, it combined readily with many elements to form oxides. As a result of this process, most of our useful metals exist as oxides in ore form. Iron is obtained from the oxides of hematite (Fe_2O_3) and magnetite (Fe_3O_4). Sand is an oxide of silicon (SiO_2). During the molten state, hydrogen and oxygen combined to form steam. When *coalescence* occurred, the steam condensed to form water to fill basins and oceans. What is this oxide called? How many other oxides can you name?

Acids. Acids are compounds that always contain hydrogen. Acids have a sour taste; their water solution turns blue litmus paper red and methyl orange to pink.

Bases. Bases are compounds that produce OH or hydroxyl ions in solution and that react with and neutralize acids. They always have at least one atom of oxygen and one atom of hydrogen that have united to form a hydroxide, OH. We learned from the discussion of acids that hydrogen is always present in an acid. Likewise, the OH, or hydroxyl ion, is always a part of the base. Also, bases affect indicators, such as litmus paper and phenolphthalein, too. The water solution of a base turns red litmus paper blue, and phenolphthalein turns to a deep red color. Another indicator, Congo red, will turn to a scarlet color in the presence of bases, while acids will turn it to blue. Most bases have a bitter taste.

Salts. When acids and bases react with one another, a *salt* and water are formed. The salt which is produced contains a positive metallic ion other than hydrogen and a negative nonmetallic ion other than the hydroxyl. Since salts do not contain a common ion like acids and bases, they differ widely in their properties. Some salts have a sour taste, while others are bitter or just salty. Many dissolve easily in water, and others are only slightly soluble or do not dissolve at all.

Since many salts are soluble, they are dissolved in surface waters of streams. These streams cross mountains, plateaus, and plains as they find their way toward the sea. There the salts in suspension are deposited in layers of sediment on ocean bottoms and remain permanently, causing the oceans to be salty.

On the other hand, some salts are insoluble. Clay, for example, is chiefly composed of a salt called hydrogen aluminum silicate and is used to make bricks. The marble and limestone in buildings are largely salts of calcium carbonate. Many of our ores are useful salts. Lead is obtained from the salt, lead sulfide, which is known as galena. Zinc comes from zinc sulfide or zinc blende.

TYPES OF CHEMICAL REACTIONS

Each rock or mineral may be thought of as a chemical system in which certain chemical changes take place. It is generally agreed among scientists that the earth's crust is far different now from what it was in the beginning. It is necessary, then, for us to become aware of the changes in the earth's crust. To do so, we must understand fully how the elements and compounds react to produce these chemical changes. Soil science is chiefly concerned with the reactions between the constituents of the earth itself, the reactions caused by water, and the reactions produced by the agency of the atmosphere.

Equations. Scientists use a system of shorthand, so to speak, in order to show by symbols what actually happens when substances react with one another. In a chemical reaction, the plus sign (+) on the left of an equation means that the two elements or substances react with one another. The arrow ($\rightarrow$) means "yields" and the plus sign (+) on the right side may be read as "and." Thus, the equation $Zn + 2HCl \longrightarrow ZnCl_2 + H_2\uparrow$ is read as follows: Zinc reacts with hydrochloric acid to produce zinc chloride and hydrogen. The vertical arrow after hydrogen indicates that it is a gas. The equation for the reaction of potassium hydroxide with hydrochloric acid $KOH + HCl \longrightarrow H_2O + KCl$ may be read: Potassium hydroxide reacts with hydrochloric acid to yield water and potassium chloride.

In general, chemists group chemical reactions into four common types:

1. *Simple union* or *direct combination of elements.* In a sense this reaction may be called a *synthesis,* for different elements combine to form a compound. When sulfur is burned in the presence of oxygen, a new compound, sulfur dioxide, is formed. Likewise, when hydrogen burns, hydrogen oxide or water is the resulting product. Aluminum unites directly with oxygen to form aluminum oxide. Carbon dioxide is another example of a compound formed from the direct union or combination of carbon and oxygen.

It is necessary for us to become familiar with the process of writing equations in chemical reactions. For our purpose in this course, we do not have to know all the rules that chemistry teachers use. We do need to know, however, certain simple rules and steps involved. Consider the example in the paragraph above, where hydrogen burns with oxygen.

The *first* step is to write the word equation for the burning process. This reaction is as follows: Hydrogen + oxygen $\longrightarrow$ hydrogen oxide (water).

Then the *second* step is to write the correct symbol or formula for each of the two substances which react with one another:

$$H_2 \quad + \quad O_2 \longrightarrow$$

Notice that hydrogen and oxygen are listed as H_2 and O_2 because they are molecules. (A molecule is the smallest unit quantity of matter that can exist by itself in the gaseous state. The hydrogen and oxygen molecules are composed of two atoms.)

The *third* step is to write the formula for the compound which has been formed. We know that hydrogen and oxygen will combine in some manner.

$$H_2 \quad + \quad O_2 \longrightarrow HO$$

How do we know what the exact combination will be? From our study of valence, recall that electrons are involved when two or more elements join to form a compound. Now look at the periodic table on page 292 and observe that hydrogen has a valence of one and oxygen two. Place the one above hydrogen and two above oxygen.

$$H_2 \quad + \quad O_2 \longrightarrow \overset{12}{HO}$$

The lowest common multiple of 1 and 2 is two. Place this in the position as follows:

$$H_2 \quad + \quad O_2 \longrightarrow \overset{\overset{2}{\wedge}}{\underset{}{12}} HO$$

Divide and the result is H_2O.

$H_2 + O_2 \longrightarrow H_2O$. This means that hydrogen has two electrons to give away and oxygen receives two electrons from the two hydrogen atoms.

The *fourth* step is to balance the equation. To be balanced the entire equation must have the same number of atoms on the left-hand side of the arrow as it does on the right-hand side. After we have written the correct formula of substances which have reacted, we can change the number of molecules in any way we see fit in order to balance the equation. The correct equation now reads:

$$2H_2 \quad + \quad O_2 \longrightarrow 2H_2O$$

It is easy to see that there are four hydrogen atoms and two oxygen atoms on each side of the arrow. It is now balanced. Can we use the same procedure to explain the equation which represents a union of aluminum and oxygen?

$$4Al \quad + \quad 3O_2 \longrightarrow 2Al_2O_3$$

Use the valence tables found in the *periodic table*.

2. *Simple decomposition* of a compound may take place due to the influence of heat, light, or electricity. In this case, the compound is broken up into its constituents. For example, when an electric current is passed through water, the water is decomposed into the gases, hydrogen and oxygen. This process is known as *electrolysis*.

$$2H_2O \longrightarrow 2H_2\uparrow \quad + \quad O_2\uparrow$$

Ammonium hydroxide (NH_4OH) is very unstable and when it is heated gently, it decomposes into ammonia (NH_3) and water (H_2O).

$$NH_4OH \longrightarrow NH_3\uparrow \quad + \quad H_2O$$

The familiar home disinfectant, hydrogen peroxide, must be kept in colored bottles because when light strikes it, it is decomposed into its component parts, water and oxygen.

$$2H_2O_2 \longrightarrow 2H_2O \quad + \quad O_2\uparrow$$

Another interesting example of decomposition is found in limestone caverns. Water containing carbon dioxide (this is a weak solution of carbonic acid) seeps slowly through layers of limestone rock (calcium carbonate).

$$H_2O \quad + \quad CO_2 \longrightarrow H_2CO_3 \quad \text{(Carbonic Acid)}$$

$$H_2CO_3 \quad + \quad CaCO_3 \longrightarrow Ca(HCO_3)_2$$
$$\text{(Calcium Bicarbonate)}$$

The calcium bicarbonate which is formed is in a solution. When this solution drips into a cave or cavern, the water evaporates, leaving the calcium carbonate in various forms. The forms are known as *stalactites* if they hang like icicles from the ceiling, and as *stalagmites* if they grow upward from the floor. If the stalactites and stalagmites meet, a pillar is formed.

$$Ca(HCO_3)_2 \longrightarrow CaCO_3 + \overset{\text{Evaporates}}{H_2O\uparrow} + \overset{\text{Gas}}{CO_2\uparrow}$$

3. *Simple displacement* is the displacement of one element in a compound by another. When a strip of aluminum foil is placed in a solution of silver sulfate, the aluminum quickly displaces the silver from the sulfate ion.

$$2Al + 3Ag_2SO_4 \longrightarrow Al_2(SO_4)_3 + 6Ag$$

Metals displace hydrogen in acids. For example, when metallic zinc is placed in hydrochloric acid, bubbles of hydrogen may be seen as they escape. Zinc replaces the hydrogen and unites with the chlorine to form zinc chloride.

$$Zn + 2HCl \longrightarrow ZnCl_2 + \overset{\text{Gas}}{H_2\uparrow}$$

4. In *double displacement* or *metathesis* two compounds exchange parts to form new compounds. A typical example of double displacement can be seen when silver nitrate and sodium chloride are placed in the same solution. Silver and sodium exchange negative ions and form new compounds, sodium nitrate and silver chloride.

$$AgNO_3 + NaCl \longrightarrow AgCl + NaNO_3$$

From the four types of chemical reactions described above, we can see that in order for reactions to take place there must be an exchange of electrons. In the study of earth and the earth's crust, we note that all the reactions may be found in nature. They take place, of course, under various conditions for nature is not static. Temperatures vary from season to season; changes take place in atmospheric pressure and humidity; and some reactions occur in solution while others do not.

THE PERIODIC CLASSIFICATION OF ELEMENTS

In 1869 the Russian chemist Dmitri Mendeleeff and the German Lothar Meyer, working independently, discovered an existing relationship between the atomic weights of elements and their physical properties and chemical activities. Subsequently, on these bases, they arranged the then known elements in the order of increasing atomic weights. This arrangement was later stated as the *periodic law*. Afterward, when the nuclear charge of the atom was discovered, it was found that the *atomic number* provided a more accurate basis for the classification of elements than the original statement of Mendeleeff and Meyer. Actually, either basis can be used because, with the exception of potassium, nickel, and iodine, the atomic weights of elements follow the same order as the atomic numbers.

The *Periodic Table* of the classification of elements shows the elements arranged in the order of their atomic weights and also places the elements with similar properties in the same column. For example, those elements with similar properties and chemical behavior are organized in such a way that they appear in vertical columns and are called *groups* or *families*. Prior to about 1898 there were eight vertical columns, but with the discovery of the noble gases, helium, neon, argon, krypton, xenon, and radon, an additional vertical column was added and referred to as the zero group.

Mendeleeff's table, as we know it, is also divided into rows across the table from left to right. These horizontal rows are called *periods*. Using the *noble gases* as reference points, the student can observe that elements which are placed between any two (e.g., lithium to fluorine, or sodium to chlorine) would comprise a *period*. No two elements in the same period are exactly alike because many different properties may be

observed. All periods are alike though, in one respect; they
end with a *noble gas*.

Two horizontal rows of elements are found at the bottom
of the *Periodic Table*. The first row, elements 58–71 (cerium
through lutetium), is known as the lanthanide series and should
be placed in the proper order by atomic number. The sixth
period would then have its required 32 elements. Likewise, the
next row, elements 90–103 (thorium through lawrencium), is
called the actinide series and should be placed in the seventh
period, which is now incomplete. For convenience, they are
usually omitted from the table proper and placed in groups
by themselves at the bottom of the table.

THE ELEMENTS

*Atomic Weights are based on Carbon 12.
**Mass numbers of isotopes of elements with the longest half
life periods are in parentheses.

	SYMBOL	ATOMIC NUMBER	ATOMIC WEIGHT
Actinium	Ac	89	(227)
Aluminum	Al	13	26.9815
Americium	Am	95	(243)
Antimony	Sb	51	121.75
Argon	Ar	18	39.948
Arsenic	As	33	74.9216
Astatine	At	85	(210)
Barium	Ba	56	137.34
Berkelium	Bk	97	(247)
Beryllium	Be	4	9.0122
Bismuth	Bi	83	208.980
Boron	B	5	10.811
Bromine	Br	35	79.909
Cadmium	Cd	48	112.40
Calcium	Ca	20	40.08
Californium	Cf	98	(251)
Carbon	C	6	12.01115
Cerium	Ce	58	140.12
Cesium	Cs	55	132.905
Chlorine	Cl	17	35.453
Chromium	Cr	24	51.996
Cobalt	Co	27	58.9332
Copper	Cu	29	63.54

	SYMBOL	ATOMIC NUMBER	ATOMIC WEIGHT
Curium	Cm	96	(247)
Dysprosium	Dy	66	162.50
Einsteinium	Es	99	(254)
Erbium	Er	68	167.26
Europium	Eu	63	151.96
Fermium	Fm	100	(253)
Fluorine	F	9	18.9984
Francium	Fr	87	(223)
Gadolinium	Gd	64	157.25
Gallium	Ga	31	69.72
Germanium	Ge	32	72.59
Gold	Au	79	196.967
Hafnium	Hf	72	178.49
Helium	He	2	4.0026
Holmium	Ho	67	164.930
Hydrogen	H	1	1.00797
Indium	In	49	114.82
Iodine	I	53	126.9044
Iridium	Ir	77	192.2
Iron	Fe	26	55.847
Krypton	Kr	36	83.80
Lanthanum	La	57	138.91
Lawrencium	Lw	103	(257)
Lead	Pb	82	207.19
Lithium	Li	3	6.939
Lutetium	Lu	71	174.97
Magnesium	Mg	12	24.312
Manganese	Mn	25	54.9380
Mendelevium	Md	101	(256)
Mercury	Hg	80	200.59
Molybdenum	Mo	42	95.94
Neodymium	Nd	60	144.24
Neon	Ne	10	20.183
Neptunium	Np	93	(237)
Nickel	Ni	28	58.71
Niobium	Nb	41	92.906
Nitrogen	N	7	14.0067
Nobelium	No	102	(253)
Osmium	Os	76	190.2
Oxygen	O	8	15.9994
Palladium	Pd	46	106.4
Phosphorus	P	15	30.9738
Platinum	Pt	78	195.09
Plutonium	Pu	94	(244)
Polonium	Po	84	(209)
Potassium	K	19	39.102
Praseodymium	Pr	59	140.907
Promethium	Pm	61	(145)
Protactinium	Pa	91	(231)
Radium	Ra	88	(226)

	SYMBOL	ATOMIC NUMBER	ATOMIC WEIGHT
Radon	Rn	86	(222)
Rhenium	Re	75	186.2
Rhodium	Rh	45	102.905
Rubidium	Rb	37	85.47
Ruthenium	Ru	44	101.07
Samarium	Sm	62	150.35
Scandium	Sc	21	44.956
Selenium	Se	34	78.96
Silicon	Si	14	28.086
Silver	Ag	47	107.870
Sodium	Na	11	22.9898
Strontium	Sr	38	87.62
Sulfur	S	16	32.064
Tantalum	Ta	73	180.948
Technetium	Tc	43	(97)
Tellurium	Te	52	127.60
Terbium	Tb	65	158.924
Thallium	Tl	81	204.37
Thorium	Th	90	232.038
Thulium	Tm	69	168.934
Tin	Sn	50	118.69
Titanium	Ti	22	47.90
Tungsten	W	74	183.85
Uranium	U	92	238.03
Vanadium	V	23	50.942
Xenon	Xe	54	131 30
Ytterbium	Yb	70	173.04
Yttrium	Y	39	88.905
Zinc	Zn	30	65.37
Zirconium	Zr	40	91.22

Word List for Study

mass	combustion	atomic number
weight	valence	proton
element	chemical change	electron
compound	physical property	mixture
isotope	electrolysis	chemical equation
stalactite	periodic table	atomic weight
stalagmite	synthesis	matter
neutron	atom	chemical property
ion	solution	filtration
molecule	atomic mass number	coalescence

Biographies to Investigate

Joseph Priestley   Ernest Solvay

Dmitri Mendeleeff   Sir Joseph John Thomson

Robert A. Millikan   Ernest Rutherford

Niels Bohr   James Chadwick

Antoine Lavoisier   Harold C. Urey

Lothar Meyer

Questions for Consideration

1. What is the significance of the electrons in the outer orbits of atoms?
2. How can you distinguish between acids, bases, and salts?
3. What is an oxide?
4. Select six atoms that we have studied in this chapter. Refer to the periodic table, p. 292. Diagram the atoms showing the protons, electrons, and neutrons.
5. Distinguish between a metal and a nonmetal.
6. Write the word equation for the reaction of iron with sulfuric acid. Give the formula equation and balance it.
7. Can you tell the difference between a mixture and a compound? Give some examples.
8. What is meant by a balanced equation?
9. Write the equation for the burning of hydrogen and oxygen.
10. What is an example of the formation of calcium carbonate in nature?
11. Give an equation that shows the commercial preparation of hydrogen. Use the electrolysis of water.

12. What element does an acid always contain?
13. Distinguish between the physical properties and the chemical properties of a substance.
14. What causes iron to rust? Why does silver tarnish?
15. Why are there so many oxides in the earth's crust?
16. Find out what percentage of the earth's crust is oxygen.

Additional Activities

1. Dissolve a given quantity of salt in water and taste it. Let the solution stand until the class period tomorrow. Compare the taste. Explain the difference, if any.
2. Write a word equation and a formula equation for each of the reactions listed below. What type of reaction is it? Use the periodic table, p. 292.
   - a. Carbon + oxygen
   - b. Hydrogen + oxygen
   - c. Zinc + hydrochloric acid
   - d. Silver + sulfur
   - e. Potassium + water
   - f. Ammonium hydroxide + heat
   - g. Barium chloride + sodium sulfate
   - h. Water + carbon dioxide
   - i. Carbonic acid + heat
   - j. Hydrogen peroxide + light
   - k. Water + electricity
   - l. Aluminum + silver sulfate
   - m. Limestone + carbonic acid
   - n. Aluminum + oxygen
   - o. Sodium iodide + chlorine
3. Explain why minerals are present in water.
4. Find out the names of several of the great limestone caves in the United States. Describe how these caves were formed.

5. Write a short paper on the contributions of Joseph Priestley to modern civilization.

6. If hydrochloric acid and sodium hydroxide were mixed together in proper proportions, would it be harmful to drink them? Write the equation and explain the reaction.

7. Why does sea water contain many salts in solution? How do the salts get into the oceans?

8. Why are some salts soluble and others insoluble? What part does insolubility play in the mining industry?

9. Find information with which to devise an experiment that will prove that ions exist in a solution.

10. Name as many indicators as you can and explain how they act in acids and bases.

11. Find out why silver bromide salt must be produced in a darkroom.

APPENDIX B

Soils of North Carolina

SMALL ACREAGE SUPPORTS GREAT PRODUCTION

North Carolina enjoys a relatively high position in agriculture. It ranks among the top ten states in total farm income and fifth in income from crops. Yet less than one-fourth of the total land area of the state is devoted strictly to agricultural use. Its cropland area is nearly three times that of its pasture acreage, but much cropland is idle, fallow, or otherwise nonproductive. In 1964, crops were harvested from only one acre in eight of North Carolina's total land area. The amount of crop-producing land has decreased greatly in recent years. This decline in the harvested-crop acreage continues, while at the same time, the total crop production increases steadily. Thus the high productive qualities of North Carolina's agricultural soils demonstrate the great value of its soil resources.

SOIL REGIONS

North Carolina's soils are varied and complex because of the great variety of geologic conditions, relief features, and natural drainage within the state. The most important properties that distinguish North Carolina soils from each other are parent material (derived from rocks), slope, and drainage. The three major geographic regions of North Carolina, the Coastal Plain, the Piedmont, and the Mountains, differ from each other in these soil properties, so that the geographical regions may be considered soils regions, also.

GENERALIZED SOILS

SOIL GROUPS ASSOCIATED WITH MOUNTAINOUS AREAS

Approximate land area, 5,186,080 acres, 16.5 per cent of state

Miscellaneous alluvium, terrace soils, stony rough land—1,360,000 acres
Soil Associations
Stony Rough Land, 1,300,000 a.; Alluvial Soils, 50,000 a.; Terrace Soils, 10,000 a.

Upland soils derived from acidic crystalline rocks, chiefly granites, gneisses, and schists—3,082,880 acres
Soil Associations
Porters-Ashe, 1,988,880 a.; Halewood-Hayesville, 760,000 a.; Clifton-Porters, 270,000 a.; Edneyville-Balfour, 64,000 a.

Upland soils derived from mica-rich schistose rocks—760,000 acres
Soil Associations
Fannin-Watauga, 110,000 a.; Talladega-Ramsey, 650,000 a.

SOIL GROUPS ASSOCIATED WITH THE PIEDMONT

Approximate land area, 12,200,000 acres, 38.8 per cent of state

Upland soils derived from acidic or mixed acidic and basic crystalline rocks —7,100,000 acres
Soil Associations
Appling-Cecil, 2,220,000 a.; Cecil-Lloyd, 2,200,000 a.; Hayesville-Cecil, 1,040,000 a.; Helena-Wilkes, 1,010,000 a.; Madison-Surry, 630,000 a.

Upland soils derived from "Carolina Slates"—1,740,000 acres
Soil Associations
Alamance-Orange, 750,000 a.; Herndon-Georgeville, 750,000 a.; Georgeville-Tirzah, 240,000 a.

Upland soils derived from basic crystalline rocks—1,000,000 acres
Soil Associations
Davidson-Mecklenburg, 520,000 a.; Iredell-Mecklenburg, 480,000 a.

Upland soils derived from Triassic sediments—850,000 acres
Soil Associations
Mayodan-Granville, 360,000 a.; White Store-Creedmoor, 310,000 a.; Granville-White Store, 180,000 a.

SOIL GROUPS ASSOCIATED WITH THE COASTAL PLAIN

Approximate land area, 14,036,000 acres, 44.7 per cent of state

Light textured, light colored, well drained to somewhat poorly drained—6,771,000 acres
Soil Associations
Norfolk-Ruston, 2,874,000 a.; Dunbar-Lynchburg, 1,900,000 a.; Craven-Shubuta, 870,000 a.; Lynchburg-Rains, 750,000 a.; Dragston-Fallsington, 225,000 a.; Terrace Soils, 150,000 a.

Light textured, light colored, excessively drained
Soil Associations
Lakeland-Norfolk, 1,300,000 a.

Dark colored, poorly drained soils, muck and swamp—4,380,000 acres
Soil Associations
Muck-Peat, 1,360,000 a.; Lenoir-Coxville, 630,000 a.; Bladen-Elkton, 600,000 a.; Swamp-Tidal Marsh, 505,000 a.; Coxville-Bladen, 450,000 a.; Rutlege-Plummer, 350,000 a.; Portsmouth-Hyde, 315,000 a.; Bayboro-Muck, 170,000 a.

Miscellaneous, hardpan, alluvium, and dunesand—1,585,000 acres
Soil Associations
Kle-Leon (hardpan), 1,450,000 a.; Alluvial Soils, 30,000 a.; Coastal Beach-Dunesand, 105,000 a.

Modified from W. D. Lee and E. F. Goldston

Major regional soil distinctions. A technical (1949) classification of North Carolina soils shows 12 great soil groups, 38 soil associations, and over 200 soil series. The better-drained, mature soils belong largely to the group known as the *Red-Yellow Podzolic* group that covers much of the southeastern United States. (See map, page 78 and table 2, p. 75.) In North Carolina, this group of soils comprises about three-fourths of the soils of the Piedmont and one-half those of the Coastal Plain. In the Mountains, because of lower temperatures resulting from higher elevations, *Gray-Brown Podzolic* soils are common. There are also a few areas of Podzols. These two soil groups are characteristic of northeastern United States. About one-seventh of the state's soils are poorly drained. They are found largely in the eastern part of the Coastal Plain. These wet soils are referred to by such technical names as *Bog, Low Humic Gley, Ground-Water Podzols,* and other terms that suggest poor drainage. Most North Carolina soils are relatively low in plant nutrients and are mildly to strongly acid. Thus heavy fertilization and liming are common practices.

Of the three soil properties, parent material, slope, and drainage, drainage is of greatest significance in the Coastal Plain. In the Piedmont, where most soils are well drained, rock conditions provide the major basis for classification. Steep slope and increased elevation account largely for soil characteristics of the Mountains.

COASTAL PLAIN SOILS ARE GROUPED ACCORDING TO DRAINAGE

The inner edge of the Coastal Plain varies in elevation from 100 to 300 feet above sea level except in the Sand Hills of the southwestern part, where the altitude ranges from 300 to 650 feet. Drainage is thorough throughout the belt. In general, the soils tend to be light in color, but they vary from one area to another according to parent material and drainage.

Drainage is the dominant factor in classifying and using

Coastal Plain soils. The accompanying North Carolina map shows soils of the Coastal Plain region arranged in four groups according to drainage and related factors. Light-textured, light-colored, well-drained to somewhat poorly drained soils dominate the inner Coastal Plain and cover nearly one-half of the entire region (Group No. 1). Soils are predominantly well-drained sandy loams and are highly prized for tobacco, cotton, peanuts, and corn. They are among the most valuable soils of the state. The Norfolk, Ruston, and Marlboro series described later in this appendix (CP–8, 10, and 7) typify those excellent soils.

Excessively drained soils (Group No. 2) are distinctly sandy and are overly drained. They are found chiefly in the Sand Hills area. These soils are inferior for most crops, but they support thriving peach orchards.

Dark-colored, poorly drained soils, muck, and swamp (Group No. 3) comprise one-third of the total Coastal Plain area. These soils lie mainly near the coast, surrounding the bays and sounds and extending far up the major streams. Artificial drainage has brought much of this land into crop production. The Portsmouth and Hyde series (CP–9 and 3) and the Bladen series are typical of these poorly drained soils.

Soils not dominated by drainage problems. A miscellaneous group of soils (Group No. 4) covers about one-tenth of the Coastal Plain. This group is dominated by hardpan soils, of which the Leon series (CP–4) is a good example. The most extensive hardpan areas are in the southern corner of the state. Agricultural use of these soils is limited. Most of the blueberries of the region are grown on these hardpan soils. Coastal beach, dunes, and river alluvium account for only a limited acreage in this miscellaneous group.

PIEDMONT SOILS ARE GROUPED ACCORDING TO GEOLOGICAL CONTRASTS

Four classes of rocks—acid crystallines, basic crystallines, "Carolina Slates," and Triassic sediments—give rise to four major groups of well-drained Piedmont soils. A miscellaneous group embraces alluvium, terrace soils, and stony rough land.

Soils derived from acid or mixed acid and basic crystalline rocks (Group No. 5) occupy nearly three-fifths of the Piedmont. They are the major soils of two large areas. One area covers the northeastern corner of the region. The other area spans the entire Piedmont in its western half. Typical of this group are the red upland soils derived from acid gneisses, schists, and granites. These red soils are characteristic of the Appalachian Piedmont from Maryland to Georgia. One of the most widely distributed and best known of these red soils is the Cecil series (P-2). These soils tend to have sandy loam surfaces and clay loam to clay subsoils. Mixed with the red soils and often occupying rolling to flattish upland areas are gray soils with yellow or red subsoils. Two important soils in this category are the Appling and the Durham series (P-1 and 5). This group contains some of the best farm land in the Piedmont, and it is well-suited to corn, small grains, lespedeza and other hay crops, cotton, tobacco, and pasture. Erosion is advanced, however, and difficult to control. A large proportion of this land is in timber.

Soils derived from "Carolina Slates" (Group No. 6) span the east-central part of the region, occupying one-seventh of the Piedmont. They resemble the group of soils just discussed (Group No. 5) in color and texture; they are also similarly used, except for the fact that cotton and tobacco are less often grown in slate soils. The Georgeville series (P-6) is a good example.

Distributed widely over the Piedmont but more prevalent in the western half are relatively heavy-textured red and brownish-red soils. They are derived from dark-colored basic crystalline rocks (Group No. 7). These soils, which comprise about one-twelfth of the Piedmont acreage, for the most part are relatively small in area and exhibit a patchy pattern. This group contains the Davidson and Mecklenburg series (p-4 and 9), which are among the best soils of the Piedmont for most crops except tobacco. Beef cattle and dairy herds do well on these soils.

The Durham-Sanford basin and other smaller areas lying largely in the eastern edge of the Piedmont consist of sand-stones and shales of the Triassic period. Soils derived from these rocks (Group No. 8) cover only 7% of the Piedmont. They have light-colored sandy loam surfaces with clay loam to plastic clay subsoils. With slow internal drainage because of tight subsoils, these soils are of only medium quality for general crops, but some are well suited to tobacco. The White Store (P-10), Creedmore, Wadesboro, and Mayodan series are in this group.

Miscellaneous Piedmont soils. A miscellaneous group of Piedmont soils comprising about one-tenth of the total area is represented on the map as Group No. 9. River alluvium and river terrace soils make up most of this acreage. A small part of it consists of stony, rough land without true soil pro-file features. Much of the alluvial acreage (for example, the Congaree Series, P-3) is hampered by wetness and frequent flooding. Most of these soils are still covered with timber. Some of them, however, are planted in shallow-rooted crops, includ-ing corn and hay. Frequently, good yields are harvested from such soils if they escape summer and autumn floods. Terrace soils (Wickham Series, P-11) are among the best crop soils of the Piedmont, though they comprise less than 3% of the total acreage. Most Piedmont terrace soils are under cultivation.

SOILS OF THE MOUNTAINS

Due to the high elevations and the lower temperatures that result from them, many soils of the Mountains are identi-cal with the Gray-Brown Podzolic and Podzol soils commonly found further north. Acid crystalline rocks provide the parent material for three-fifths of the soils of this region (Group No. 10). These brown-clay or clay-loam soils usually occur on relatively steep slopes and are stony and shallow (Porters and Ashe series, M-3 and 1). Less than one-tenth of this acreage is used for crops, which consist largely of corn, hay, cabbage, and Burley tobacco. Other mountain soils derived from micaceous schistose rocks make up about one-sixth of the total acreage (Group No. 11). These soils, of which the Ramsey and Avery series (M-4 and 2) are examples, are stony, shallow, steep, and of slight agricultural importance.

One-fourth of the mountain area consists of stony, rough land (Group No. 9). These soils are of virtually no agricultural use. A limited acreage of alluvial and terrace soils is included in Group No. 9. They are intensively used for crops and pasture, and they are highly prized.

REPRESENTATIVE NORTH CAROLINA SOIL SERIES

COASTAL PLAIN

CP-1. COXVILLE SERIES

Classification

Low-Humic Gley (1949 classification): Aquult (Subord. 7th. Approx.); Ochraquult (Grt. Group, 7th. Approx.)

General Aspects

The dark gray, fine-textured, poorly drained Coxville soil series is widely distributed in the lower and middle Coastal Plain from Virginia to Florida and in Alabama and Mississippi. It comprises about 1½% of North Carolina's Coastal Plain acreage. It is submaturely developed and is limited to nearly level topography with slopes under 2%. It has slow external and internal drainage, and is strongly acid (pH 5–5.5).

Profile Features

A-horizon (0–12 in.): Gray to very dark gray sandy loam, loam, or silt loam.

B-horizon (1–2 ft. thick): Mottled with gray, yellow, pale olive, or pale brown, often splotched with red; fine sand to clay, becoming heavier with depth; very firm when moist, plastic when wet, and hard when dry.

Utilization

Coxville soil requires artificial drainage, heavy liming, and fertilization for crops. Corn, some cotton, soy beans, truck crops, and strawberries are grown. A large proportion of the Coxville soil is devoted to timber.

CP-2. DUNBAR SERIES

Classification

Low-Humic-Gley (1949 classification): Ochrult (Subord. 7th. Approx.); Normochrult (Grt. Group, 7th. Approx.)

General Aspects

The gray to grayish-brown, fine-textured, somewhat poorly drained Dunbar soil series is widely distributed in all parts of the Coastal Plain from Virginia to Florida and in Alabama and Mississippi. It comprises about 3% of North Carolina's Coastal Plain acreage. It is a deep soil and is better drained than the Coxville but not as well drained as the Norfolk series. It is limited to a nearly level topography with slopes sometimes up to 4%. It is strongly to moderately acid (pH 5–5.5).

Profile Features

A-horizon (0–12+ in.): Gray to grayish-brown (depending upon organic matter), loose sandy loam to friable, very fine sandy loam.

B-horizon (6–30+ in. thick): Lightly mottled with gray, yellow, and sometimes red; firm sandy clay loam to very fine sandy clay; weak, blocky structure; slightly plastic when wet, slightly hard when dry.

Utilization

Dunbar soil is good for most crops except truck crops. More than one-half of it has been cleared for crops in some North Carolina counties. It requires some artificial drainage, liming, and fertilizing; its chief crops are cotton, corn, and peanuts. Crops hold up well in drier-than-average years.

CP-3. HYDE SERIES

Classification

Humic-Gley (1949 classification): Aquept (Subord. 7th. Approx.); Umbraquept (Grt. Group, 7th. Approx.)

General Aspects

It is dark gray to black, finely textured, very poorly drained, strongly to moderately acid (pH 5–5.5). This Hyde soil series is distributed over the lower Coastal Plain from Virginia to Florida. It makes up over 1% of North Carolina's Coastal Plain acreage. Hyde loam is one of the wettest of the Coastal Plain soils, grading into muck in many places. It is limited to a nearly flat topography, with slopes under 2%.

Profile Features

A-horizon (18–30 in. thick): Very dark gray to black loam, high in organic matter.

B-horizon (thick, poorly defined; grades into C Gley layer): Very dark gray or grayish-brown loam to fine sandy clay: high in organic matter; slightly plastic when wet and hard when dry.

Utilization

In some counties as much as 90% of the Hyde soil is forested, for it is of little value unless it is drained, limed, and fertilized. But if these things are done, it becomes very productive and yields good corn, truck, soy bean, and hay crops.

CP-4. LEON SERIES

Classification

Ground-Water Podzol (1949 classification): Aquod (Subord. 7th. Approx.); Humaquod (Grt. Group, 7th. Approx.)

General Aspects

This light-colored ("pepper and salt"), light-textured, somewhat poorly drained, hardpan Leon soil is widely found in the lower Coastal Plain from New Jersey to Alabama but its largest acreages lie in Georgia and Florida. It is largely limited to level topography with slope under 2%. It comprises about 1½% of North Carolina's Coastal Plain acreage. It is strongly to moderately acid (pH 5–5.5).

Profile Features

A-horizon (3–12 in. thick): Gray, loose sand surface with very light gray, yellowish-white, or white loose sand underneath.

B-horizon (poorly defined; characterized by pan layer): Dark brown to brownish-black, firm to very dense, organic hardpan 2–6 inches thick, usually at depths of 10 to 30 inches; pan layer is underlain .by brownish-gray, gray, or yellowish-brown loose sand.

Utilization

Leon soil is of very little agricultural value, though blueberries grow well on it in North Carolina and a limited acreage of it is devoted to truck crops. But most Leon soil is left to timber growing.

CP-5. LYNCHBURG SERIES

Classification

Low-Humic-Gley (1949 classification): Ochrult (Subord. 7th. Approx.); Normochrult (Grt. Group, 7th. Approx.)

General Aspects

The deep, gray to pale yellow, light-textured, somewhat poorly drained Lynchburg soil series is distributed through all parts of the Coastal Plain from North Carolina to Mississippi. It is one of North Carolina's leading Coastal Plain soils in acreage, comprising about 6% of the total. It is one of the better drained Low-Humic-Gley series, occupying slopes up to 5%. It is strongly acid (pH about 5.2).

Profile Features

A-horizon (6–18 in. thick): Gray to dark gray sand to loamy sand, shading into gray, friable sandy loam.

*B-horiz*on (thick, up to 3 + feet): Pale yellow loamy sand to fine sandy loam; lightly mottled with brown and gray; crumb to blocky structure.

Utilization

The Lynchburg soil requires artificial drainage, liming, and fertilizing for crop production. It handles well and is suited to a wide range of crops, including corn, cotton, soy beans, small grains, and tobacco. Yields are somewhat above the average of associated soils.

CP-6. MAGNOLIA SERIES

Classification

Red-Yellow Podzolic (1949 classification): Ochrult (Subord. 7th. Approx.); Normochrult (Grt. Group, 7th. Approx.)

General Aspects

The deep, grayish-brown to reddish-brown well-drained sandy loam soil is found in the middle and upper Coastal Plain from North Carolina to Florida, but very little of it lies in North Carolina. It occupies slopes up to 18% (usually below

7%) and often exhibits a strong reddish color where erosion and oxidation are advanced. It is strongly to moderately acid (pH 5.5 +).

Profile Features

A-horizon (0–6 + in.): Grayish-brown to reddish-brown loose sandy loam to sandy clay loam.

B-horizon (thick; up to 3 or more feet): Red sandy clay, slightly plastic when wet and hard when dry. Weak, medium subangular blocky structure.

Utilization

The Magnolia series is one of the better crop soils of the Coastal Plain. It tills well and responds well to good treatment, including liming, fertilizing, and crop rotation. It requires little drainage, but on the steeper slopes, erosion is a problem. It is suited to a variety of crops, including cotton, corn, tobacco, and hay.

CP-7. MARLBORO SERIES

Classification

Red-Yellow Podzolic (1949 classification): Udult (Subord. 7th. Approx.); Palendult (Grt. Group, 7th. Approx.)

General Aspects

This deep, gray to brownish-gray sandy loam to friable very fine sandy loam is widely distributed over the middle and upper Coastal Plain from Virginia to Florida, and in Arkansas. The Marlboro series comprises about 1½% of North Carolina's Coastal Plain acreage. It occupies relatively smooth topography with slopes up to 4%. It is moderately acid (pH 5–5.5).

Profile Features

A-horizon (0–9 in.): Gray to brownish-gray, loose sandy loam to sandy clay loam.

B-horizon (thick: 1–3 + ft.): Yellowish-brown to brownish-yellow, fine to very fine sandy clay loam; slightly plastic when wet, hard when dry; weak, subangular blocky structure.

Utilization

The Marlboro series is one of the best agricultural soils of the Coastal Plain. It handles well and responds well to good treatment, including liming, fertilizing, some draining, and crop rotation. It is highly prized for tobacco, cotton, corn, and other crops. A large proportion of Marlboro soil is in cultivation.

CP-8. NORFOLK SERIES

Classification

Red-Yellow Podzolic (1949 classification): Ochrult (Subord. 7th. Approx.); Normochrult (Grt. Group, 7th. Approx.)

General Aspects

The deep, well-drained, gray to pale yellow sand to sandy loam Norfolk series is one of the most widely distributed soils of the middle and upper Coastal Plain from Virginia to Texas. It comprises about one-twelfth of the total acreage of the North Carolina Coastal Plain. It is found in relatively level terrain, occupying slopes up to 8%, usually under 4%. It is strongly to moderately acid (pH 5–5.5).

Profile Features

A-horizon (0–12 in. thick; considerable proportion missing from some eroded slopes): Gray to pale yellow sand to sandy loam.

B-horizon (thick; up to 2–3 + ft.): Yellow to brownish-yellow sandy loam to clay loam, occasionally faintly mottled; weak to heavy blocky structure; slightly plastic when wet, hard when dry.

Utilization

The Norfolk series is the classic example of the yellow member of the Red-Yellow Podzolic group of the Atlantic-Gulf Coastal Plain. It is one of the best soils of the region. It responds to good treatment, including a minimum of drainage, heavy liming and fertilizing, and also crop rotation, particularly on steeper slopes. It is the choice soil of the region for cotton, tobacco, peanuts, and a variety of other crops. A large proportion of it is used for crops.

CP-9. PORTSMOUTH SERIES

Classification

Humic-Gley (1949 classification): Aquept (Subord. 7th. Approx.); Umbraquept (Grt. Group, 7th. Approx.)

General Aspects

This very dark gray to black, very poorly drained, sand to loamy sand soil is widely distributed in all sections of the Coastal Plain, from New Jersey to Texas. It is most commonly found on the outer Coastal Plain, often in association with marshes and swamps. It comprises about 2½% of North Carolina's Coastal Plain acreage. It usually is found on nearly flat surfaces with slopes under 2%. It is strongly to very strongly acid (pH about 5).

Profile Features

A-horizon (horizons indistinct): Very dark gray to black, loamy sand to loam surface soil; high in organic matter.

B-horizon (faintly developed): Light gray sandy loam to clay loam upper part, becoming mottled with yellow or brown below; weak blocky structure. Gley layer (C-horizon) often mottled with yellow and brown; sticky when wet, friable when moist; seldom dry; weak crumb structure.

Utilization

Portsmouth soil is not adapted to crops unless drained, limed heavily, and fertilized. Much of it is still in forests. A considerable acreage is used for pasturage. Leading crops on improved tracts include corn, soy beans, and truck crops. Tobacco, cotton, and peanuts do poorly on Portsmouth soil.

CP-10. RUSTON SERIES

Classification

Red-Yellow Podzolic (1949 classification): Ochrult (Subord. 7th. Approx.); Normochrult (Grt. Group, 7th. Approx.)

General Aspects

The deep, well-drained, reddish-yellow to reddish-brown, light-textured Ruston soil series is distributed widely over the middle and upper Coastal Plain from North Carolina to Texas. It comprises over 2% of North Carolina's Coastal Plain acreage. It is found on sloping terrain, on slopes up to 20% (mostly under 7%). It is more subject to erosion and is more oxidized than the Norfolk series. It is otherwise quite similar to the Norfolk, and at one time it was referred to as the "red Norfolk." It is strongly to moderately acid (pH 5–5.5).

Profile Features

A-horizon (shallow; often less than 6 in.): Reddish-yellow to yellowish-brown, friable, sandy loam to clay loam.

B-horizon (thick; 2 ft. or more): Reddish-yellow to strong red, friable, sandy clay loam; medium blocky structure; slightly plastic when wet, hard when dry; often stony.

Utilization

Ruston soil is suited to a wide range of upland crops—corn, cotton, tobacco, peanuts, hay—and is productive when limed and fertilized. It requires careful handling to restrict erosion, since much of it is on relatively steep slopes.

PIEDMONT

P-1. APPLING SERIES

Classification

Red-Yellow Podzolic (1949 classification): Ochrult (Subord. 7th. Approx.); Normochrult (Grt. Group, 7th. Approx.)

General Aspects

The yellowish-red, medium to fine-textured, moderately deep Appling soil series is found in areas of acid crystalline (granitic) rocks on the Piedmont from Virginia to Alabama. It is an important soil agriculturally and comprises over 5% of North Carolina's Piedmont acreage. It occupies slopes up to 30% (usually 3–12%) and erodes badly. It has no drainage problems. Often it is too stony for cultivation. It is strongly to moderately acid (pH 5–5.8).

Profile Features

A-horizon (0–12 in. thick): Gray, grayish-yellow, or grayish-brown, friable, sandy loam to clay loam, sometimes stony.

B-horizon (varying in thickness, sometimes up to 3+ ft.): Yellowish-red, becoming reddish-brown or streaked with red and yellow in lower part; tight clay, firm when moist, slightly plastic when wet, hard when dry; moderate to strong blocky structure; sometimes contains mica and sand grains.

Utilization

Appling soil is adapted to a variety of crops and to general farming practices. Corn does well, as do small grains and hay. Tobacco does moderately well in the sandier Appling soils but not so well in the tight clays and clay loams. Much of it is in pasture and much is in forests. Erosion control measures, liming, fertilizing, cover cropping, and crop rotation prove beneficial.

P-2. CECIL SERIES

Classification

Red-Yellow Podzolic (1949 classification): Ochrult (Subord. 7th. Approx.); Normochrult (Grt. Group, 7th. Approx.)

General Aspects

The red, medium to fine-textured, moderately deep Cecil soil series is derived from acid crystalline (granitic) rocks. It is the classic example of the red member of the Red-Yellow Podzolic great soil group of southeastern United States. It is widely distributed over the Piedmont from the Potomac River to Alabama. The Cecil series comprises more than one-tenth of North Carolina's Piedmont acreage. It is acid (pH 5–5.5) and strongly oxidized. It occupies steep topography with slopes up to 40% (mostly 5–15%) and is very erosive.

Profile Features

A-horizon (varying thickness up to 12 in.; often partially or completely eroded away): Gray, brownish-gray, or reddish-brown loamy sand to clay loam, occasionally stony.

B-horizon (varying in thickness; often up to 3+ ft.): Light red, often bright red to dark reddish-brown clay loam or light clay, firm when moist, plastic when wet, and hard when dry. Strongly angular to blocky structure; often contains sand grains and mica flakes.

Utilization

Cecil soil is widely used for crops, pasture, and timber. It is best adapted to general farming practices, requiring erosion-control measures, liming, fertilizing, crop rotation, and cover cropping. Corn, small grains, and hay do moderately well. Cotton grows well on it, as do fruit trees, particularly peaches. Tobacco is restricted to the sandier types of Cecil soils.

P-3. CONGAREE SERIES

Classification

Alluvial (1949 classification): Orthent (Subord. 7th. Approx.); Haplorthent (Grt. Group, 7th. Approx.)

General Aspects

This moderately well-drained to well-drained alluvial soil is widely distributed in Piedmont stream valleys from Pennsylvania to Alabama. It makes up nearly 2% of the North Carolina Piedmont acreage and also comprises a small acreage in the Mountains region. It is moderately acid (pH 5–5.5+). It occupies flat land, with slopes up to 3% but usually under 1%. It usually requires some drainage.

Profile Features

A-horizon (horizons not always distinct; vary in thickness up to a foot or more): Brown to reddish-brown, sandy to clay loam; very friable.

B-horizon (often indistinctly marked, and very thick, up to 10 ft. or more): Light yellowish-brown to reddish-brown, friable silt loam to firm clay loam. Slightly plastic when wet, seldom very hard when dry. Coarse sand and gravel frequently are found below 30–40 inches.

Utilization

Congaree is a choice soil for shallow-rooted crops—corn, hay, small grains, and soy beans. Often crops suffer from summer floods. Wetter areas often support good pastures. It is highly coveted for crops and pasture in rough areas, particularly in the Mountains.

P-4. DAVIDSON SERIES

Classification

Red-Yellow Podzolic (1949 classification): Ochrult (Subord. 7th Approx.); Rhodochrult (Grt. Group, 7th. Approx.)

General Aspects

This deep, dark red or maroon, fine-textured Davidson soil series is derived from dark-colored basic crystalline rocks. It is widely distributed over the Piedmont from Virginia to Alabama. It usually is found in relatively small areas and its total acreage is small. It occupies slopes up to 40% (usually 4–12%) and is subject to severe erosion. It is moderately to mildly acid (pH 5.5–6.5).

Profile Features

A-horizon (thickness varies, up to 12 in.): Dark brown to brownish-red, friable clay loam to firm clay.

B-horizon (thickness varies, often up to 3 ft. or more): Dark red to dusky red or purplish clay, firm when moist, plastic when wet, hard when dry; moderately fine to coarse, blocky structure. Usually weathered deeply.

Utilization

Davidson loam and clay loam are among the best crop soils of the Piedmont. This soil handles well in spite of steep slopes, but it requires considerable protection against erosion. It responds well to light liming and to fertilizing. It is, perhaps, the Piedmont's best alfalfa soil. Other crops include corn, small grains, hay crops, soy beans, and cotton. Its texture is too heavy for tobacco. A large percentage of Davidson soil is used for crops.

P-5. DURHAM SERIES

Classification

Red-Yellow Podzolic (1949 classification): Ochrult (Subord. 7th. Approx.); Normochrult (Grt. Group, 7th. Approx.)

General Aspects

This deep, yellow, fine sandy loam is widely distributed over the Piedmont from Virginia to Georgia, although its total acreage is not large. It is very limited in North Carolina. It is derived from acid crystalline (granitic) rocks and is the lightest-textured, lightest-colored of the Cecil-Appling-Durham group of associated soils. Durham soil probably developed under poor drainage conditions but is now considered well drained. It is an acid soil (pH 5–5.5). It is often found on upland flats with slopes usually under 5%.

Profile Features

A-horizon (thickness varies, up to 12+ in.): Gray, grayish-brown, or pale yellow, very friable sand or sandy loam.

B-horizon (thickness varies up to 3+ ft.): Pale yellow to brownish-yellow sandy clay, clay loam, or clay; at increased depths becomes streaked or sometimes mottled with yellow, red, or gray sandy clay loam; firm when moist, slightly plastic when wet, and slightly hard when dry; medium blocky structure.

Utilization

Durham soil is moderately good for a variety of crops, including corn, cotton, tobacco, and hay. It is not highly erodable, but it requires liming and fertilizing.

P-6. GEORGEVILLE SERIES

Classification

Red-Yellow Podzolic (1949 classification): Ochrult (Subord. 7th. Approx.); Normochrult (Grt. Group, 7th Approx.)

General Aspects

The moderately deep, red, medium to fine-textured Georgeville series is not widely distributed over the Piedmont, since it is limited to areas of "Carolina Slates." It is, however, found from Virginia to Georgia. It comprises 10% of North Carolina's Piedmont acreage. Georgeville resembles Cecil in appearance—distinctly red subsoil—but it is less friable and rarely contains sand or mica. It is acid (pH 5–5.5). Usually it is somewhat higher in available potassium than are soils derived from acid granitic rocks. It is strongly oxidized, slightly to severely eroded, and found on slopes up to 45% (usually under 17%).

Profile Features

A-horizon (usually thin—2–8 in.—and often completely eroded away): Gray, grayish-brown, to reddish-brown, silty, clay loam, occasionally gravelly or slaty.

B-horizon (varies in thickness; sometimes to 3+ ft.): Reddish-brown to red—sometimes bright red—smooth, silty clay; firm when moist, "slick" to sticky and plastic when wet, and hard when dry; medium blocky structure.

Utilization

Georgeville soil is used for crops and pasture. Much of the rougher land is in forests. It is best adapted to general farming, including livestock pasturing and raising corn, small grains, hay, and soy beans. Tobacco and peanuts do not yield well on Georgeville soil.

P-7. IREDELL SERIES

Classification

Planosol (1949 classification): Udalf (Subord. 7th. Approx.); Normudalf (Grt. Group, 7th. Approx.)

General Aspects

The gray to yellowish-brown, moderately deep, medium

to fine-textured Iredell series is derived from dark-colored basic crystalline rocks, as are the Davidson and Mecklenburg series. It is found from Virginia to Alabama, and in North Carolina it comprises 3% of the Piedmont acreage. It is moderately to slightly acid (pH 5.5–6.5), and occupies slopes up to 12% (usually under 5%).

Profile Features

A-horizon (usually rather thin—to 8–10 in.): Gray, grayish-brown, or brown surface, loose, sandy loam or friable loam, quickly grading to brown or yellowish-brown firm clay.

B-horizon (varies in thickness, to 2+ ft.): Gray to olive-brown, very firm clay; very plastic when wet, very hard when dry; cracks badly; slow internal drainage; rough, blocky structure.

Utilization

Iredell is considered a fairly good soil for a number of Piedmont crops although it is not good for tobacco. It is hard to till, and its yields are moderate. Corn, grains, and hay are the leading crops. Much of it is in pastures and much of it is in timber.

P-8. MADISON SERIES

Classification

Red-Yellow Podzolic (1949 classification): Ochrult (Subord. 7th. Approx.); Normochrult (Grt. Group, 7th. Approx.)

General Aspects

This grayish to reddish-brown, moderately deep, sandy, clay loam is derived from micaceous granitic rock and is found on the Piedmont from Virginia to Alabama. In North Carolina it comprises about 2% of the Piedmont acreage. It has a pH

of 5–5.5. It occupies slopes up to 40% (usually 6–15%).
It usually can be identified by numerous mica flakes.

Profile Features

A-horizon (thickness to 12 in.; often much is eroded
away): Grayish-brown to reddish-brown, friable, sandy loam
to clay loam, containing flat quartz, mica, and schist fragments.

B-horizon (thickness varies to 2 + ft.): Reddish-brown
to light red clay or clay loam; friable to firm when moist,
slightly hard when dry; medium blocky to platy structure,
containing much mica.

Utilization

Madison series is average to somewhat above average as
a Piedmont agricultural soil. It is adapted to general farming
and stock raising. Crops include corn, cotton, small grains,
hay, and soy beans. It is not suited to tobacco and peanuts.

P-9. MECKLENBURG SERIES

Classification

Red-Yellow Podzolic (1949 classification): Udalf (Subord. 7th. Approx.); Normudalf (Grt. Group, 7th. Approx.)

General Aspects

The dark gray to reddish-brown, moderately deep, clay
loam Mecklenburg soil series is widely distributed over the
Piedmont from Virginia to Alabama, comprising over 2% of
North Carolina's Piedmont acreage. Like the Davidson and
the Iredell series, it is derived from dark-colored basic crystalline rocks. It is moderately acid (pH 5.5–6) and occurs on
slopes up to 20% (usually 4–12%); it is less oxidized and
less red than the Davidson.

Profile Features

A-horizon (varies in thickness, up to a foot or more):
Dark gray to reddish-brown, friable clay loam.

B-horizon (thickness to 3+ ft.): Yellowish-red to red, silty clay to clay; firm when moist, plastic when wet, hard when dry; color becomes more varied—brown, red, yellow, olive, and gray—with depth; structure varies from blocky to massive.

Utilization

The Mecklenburg series is one of the Piedmont's better soils. It is suited to a variety of crops; much is in pastures. Corn, small grains, hay, and soy beans do well. Alfalfa thrives when the soil is limed. The texture is too heavy for tobacco.

P-10. WHITE STORE SERIES

Classification

Planosol (1949 classification): Ochrult (Subord. 7th. Approx.); Normochrult (Grt. Group, 7th. Approx.)

General Aspects

The moderately deep, gray to pale yellow, friable silt to sandy loam, White Store soil series is found largely in Virginia and North Carolina on sandstone and shales of the Triassic era. It comprises less than 2% of North Carolina's Piedmont acreage. It occupies flat to rolling terrain, with slopes up to 10% (usually 2–7%). It is strongly to moderately acid (pH 5–5.5). Internal drainage is slow.

Profile Features

A-horizon (relatively thick, up to 1 ft.): Gray, brownish-gray, pale yellow, or yellowish-white, silt loam to loose sand, grading quickly into plastic clay.

B-horizon (relatively thick, up to 3+ ft.): Reddish-brown to reddish-gray firm clay, becoming varicolored with grays, reds, and yellows mixed; very firm when moist, very plastic when wet, very hard when dry (very slow internal drainage); shrinks and cracks on drying out; medium angular to blocky structure; the redder the color the tougher, the more plastic, and the harder.

Utilization

White Store soil is used for a variety of crops, for pasture, and for timber growing. Corn, small grains, and hay are its most important crops. The sandy texture makes it a fair tobacco soil. Liming and fertilization are necessary.

P-11. WICKHAM SERIES

Classification

Red-Yellow Podzolic (1949 classification): Ochrult (Subord. 7th. Approx.); Normochrult (Grt. Group, 7th. Approx.)

General Aspects

The deep, brown sandy loam Wickham series is a widely distributed stream terrace soil in the Piedmont from Maryland to Alabama. Its acreage is not great; in the North Carolina Piedmont it comprises less than 1% of the total acreage. It is moderately acid (pH 5–5.8). It occurs on slopes up to 12% but usually under 7%. It is characteristic of intermediate terrace positions.

Profile Features

A-horizon (moderately thick, up to a foot or more): Grayish-brown to brown or brownish-red, friable, sandy loam, sometimes grading to firm clay loam, becoming heavier with depth.

B-horizon (thick, up to 3+ ft.): Brown to reddish-brown, very firm clay loam, silty clay, or clay; plastic when wet, hard to very hard when dry; medium blocky structure; mica flakes and sand grains often are found in it.

Utilization

A high percentage of Wickham and other terrace soils are used for crops. Their range includes corn, cotton, small grains, hay, and soy beans. Some tobacco is grown on Wickham soils. Liming and fertilizing are required.

p-12. WILKES SERIES

Classification

Lithosol (1949 classification): Ochrept (Subord. 7th. Approx.); Dystrochrept (Grt. Group, 7th. Approx.)

General Aspects

The shallow, gray to brownish-yellow, sandy clay loam is derived from a mixture of acid and basic crystalline rocks. It is widely distributed over the Piedmont from Virginia to Alabama, comprising nearly 4% of North Carolina's Piedmont acreage. It occupies steep land with slopes up to 60% (mostly 10–25%), and is highly erodable. It is strongly to moderately acid (pH 5–5.7).

Profile Features

A-horizon (thin, usually under 6 in.): Gray to brownish-yellow, sandy clay loam to clay loam.

B-horizon (thin, a few inches to 1 or 2 ft.): Mottled, streaked, or variegated with yellow, brown, and reddish-brown, sandy loam to clay; friable to firm when moist, plastic but not sticky when wet, and loose to hard when dry.

Utilization

The Wilkes soil is ranked only as from fairly good to poor for crops. Much of it is in pasture, and much of it is in forests. Corn, small grains, and hay are its leading crops. Its erodability makes it difficult to farm, but liming is effective in increasing its production.

MOUNTAIN SERIES

M-1. ASHE SERIES

Classification

Gray-Brown Podzolic (1949 classification): Ochrept (Subord. 7th. Approx.); Dystrochrept (Grt. Group, 7th. Approx.)

General Aspects

The shallow to moderately deep, yellowish-brown loam or stony loam Ashe series is found in high mountain areas of the southern Blue Ridge. It comprises about 8% of North Carolina's Mountain acreage. It is similar to soils found in northern United States at much lower elevations. It is derived from acid crystalline rocks, occupying rugged topography with slopes up to 90% (mostly 30–60%). It is strongly to moderately acid (pH 5–5.5), very erodable, and usually stony.

Profile Features

A-horizon (very thin, up to 6–10 in.): Dark-brown to yellowish-brown, very friable, gritty loam to sandy loam.

B-horizon (variable in thickness, to 2+ ft.): Brownish-yellow to yellowish-brown, sandy or gritty clay loam; friable when moist, slightly plastic when wet, soft to slightly hard when dry; weak crumb structure.

Utilization

The Ashe series is too steep and too stony for wide agricultural use. Yields are low. More is in pastures than in crops. Most of it is in forests.

M-2. AVERY SERIES

Classification

Podzol (1949 classification): Ochrept (Subord. 7th. Approx.); Dystrochrept (Grt. Group, 7th. Approx.)

General Aspects

The moderately deep, dark brown Avery loam is restricted to a very small acreage in the high mountains of western North Carolina. It occupies gently rolling to rough topography with slopes under 30% (usually 7–15%). It is moderately to strongly acid (pH 5–5.5). This Podzol is found at high elevations: it is similar to Podzols of the northern United States and southern Canada found at much lower elevation.

Profile Features

A-horizon (moderately thin, up to 8+ in.): Very dark brown, friable, stony, organic loam, underlain by light brownish-gray, friable to firm, silt loam or clay loam.

B-horizon (thickness varies, to 2+ ft.): Brown to dark brown, slightly cemented loam, becoming light yellowish-brown, somewhat streaked with gray and yellow at increased depths.

Utilization

The Avery series is agriculturally unimportant. Most of it is in forests.

M-3. PORTERS SERIES

Classification

Gray-Brown Podzolic (1949 classification): Ochrult (Subord. 7th. Approx.); Normochrult (Grt. Group, 7th. Approx.)

General Aspects

The shallow to moderately deep, brown to reddish-brown Porters series is widely distributed through the Blue Ridge region. It comprises about one-fifth of North Carolina's mountain acreage. It is derived from acid crystalline rocks, has a pH of 5–5.5, and is found in rugged topography with slopes up to 90% (usually 30–60%).

Utilization

Porters soil is too steep or too stony for extensive agricultural use. Little is in crops: relatively large acreages are in pasture but most of it is in forests.

M-4. RAMSEY SERIES

Classification

Lithosol (1949 classification): Ochrept (Subord. 7th. Approx.); Dystrochrept (Grt. Group, 7th. Approx.)

General Aspects

The shallow, brown Ramsey series is found in the southern Blue Ridge. It comprises about one-eighth of North Carolina's mountain acreage. It is derived from shales, slates, quartzites, and similar fine-grained rocks. It is strongly to moderately acid (pH 5–5.5) and occupies rugged terrain with slopes up to 90% (mostly 30–70%). It has good internal drainage, and it is not highly erodable, considering its steep slope.

Profile Features

A-horizon (usually thin—only a few inches thick): Brown or yellowish-brown, friable loam to clay loam; usually gravelly or stony.

B-horizon (Usually thin, up to 1 to 2 ft.): Brown to dark brown, slightly cemented loam to clay loam, becoming lighter in color and streaked with increased depth. Always contains some stony material.

Utilization

Ramsey soil is of limited value, agriculturally. It is too steep for crops, except in small patches. A larger acreage is in pasture but most of it is wooded.

M-5. WATAUGA SERIES

Classification

Gray-Brown Podzolic (1949 classification): Ochrult (Subord. 7th Approx.); Normochrult (Grt. Group, 7th. Approx.)

General Aspects

This moderately deep, grayish-brown loam is derived from highly micaceous rocks of the Blue Ridge province, in North Carolina, Virginia, and Georgia. Its acreage is very limited. It is found on only moderately rough topography, with slopes up to 35% (usually 8–24%). It occupies relatively smooth plateaus, broad lower mountain slopes, and intermountain areas. It has a pH of 5–5.5.

Profile Features

A-horizon (relatively thin, up to 8–10 in.): Grayish-brown to yellowish-brown, friable clay loam; occasionally contains stone or gravel.

B-horizon (thickness varies, up to 2+ ft.): Yellow to brownish-yellow, very micaceous clay loam; friable when moist, plastic when wet, slightly hard to hard when dry; blocky structure. Both surface soil and subsoil have slick or greasy feel, due to abundance of mica flakes. Also, it often contains some quartz and sand.

Utilization

Watauga soil is moderately good for general mountain crops and for pasture. Much of it is timbered.

APPENDIX C

Important Soil Series of the United States

Here are listed 163 United States soil series representative of all parts of the forty-eight conterminous states, with additional selections from Alaska, Hawaii, Panama Canal Zone, and Puerto Rico. Arrangement is by geographical areas. Each soil series is listed in the area in which it is most prominent, although in many instances a given series may be found in two or more areas. Only a brief description is given here. In Appendix B, "Soils of North Carolina," 27 of the soil series listed (as noted) are described in more detail. A complete description and discussion of most of the soils listed here can be found in county soil survey publications of the U.S. Soil Conservation Service, in *Atlas of American Agriculture,* Part III: "Soils of the United States," in the United States Department of Agriculture *Yearbook: Soils and Men* (1938), and in many state publications. There are several hundred county publications representing practically all parts of the United States. Specific counties are cited for most soil series as an aid in using these publications. The accompanying map shows representative county locations of 153 of the series listed.

For most soil series listed, the great soil group name (American classification, 1949) is given, followed by the order, suborder, great group, and subgroup names of the Seventh Approximation (1960). The following example illustrates this method of listing.

Series	Great Soil Group (1949)	7th Approximation Classification			
		Order	Suborder	Great Group	Subgroup
Norfolk	Red-Yellow Podzolic	Ultisol	Ochrult	Normochrult	Typic Normochrult

(App. B; CP-8), (Johnston Co., N.C.)

CONTENTS

Grouping by Geographical Areas showing List Numbers

Alphabetical Arrangement of Series Showing List Numbers

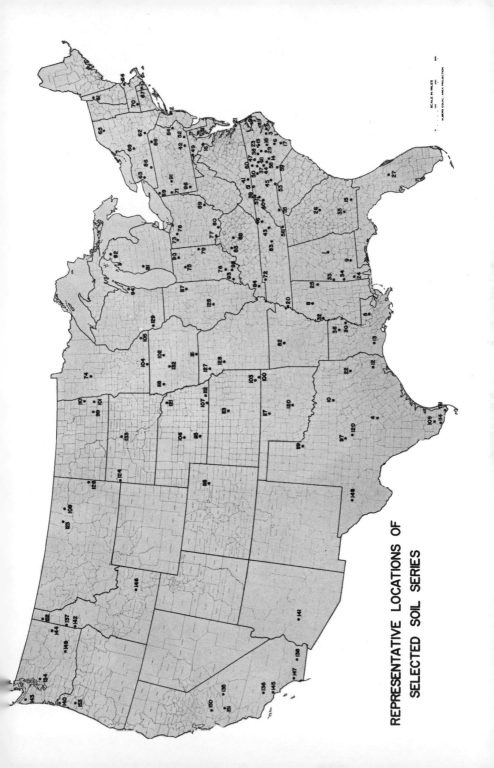

REPRESENTATIVE LOCATIONS OF
SELECTED SOIL SERIES

25. Clinton—105
26. Clyde—78
27. Colby—106
28. Collington—2
29. Congaree—44
30. Coxville—3
31. Crawford—4
32. Crete—107
33. Crosby—79
34. Crowley—5
35. Daniels—108
36. Davidson—45
37. Decatur—46
38. Dekalb—62
39. Descalabrado—160
40. Dunbar—6
41. Dunkirk—63
42. Durham—47
43. Duval—109
44. Elkton—7
45. Everett—134
46. Fairbanks—154
47. Fairmont—80
48. Fargo—110
49. Fox—81
50. Fresno—135
51. Frijoles—158
52. Georgeville—48
53. Gloucester—64
54. Greenville—9
55. Grenada—8
56. Grundy—111
57. Hagerstown—49
58. Hanceville—82
59. Hanford—136
60. Hartsells—83
61. Hastings—112
62. Hays—113
63. Helmer—137
64. Hermon—65
65. Hidalgo—114
66. Holdredge—115
67. Honouliuli—156

68. Houston—10
69. Huntington—84
70. Hyde—11
71. Imperial—138
72. Iredell—50
73. Joplin—116
74. Katy—12
75. Kirkland—117
76. Lake Charles—13
77. Lakeland—14
78. Leon—15
79. Leonardtown—16
80. Lordstown—66
81. Lowell—85
82. Lynchburg—17
83. Madera—139
84. Madison—51
85. Magnolia—18
86. Mahoning—86
87. Mamala—157
88. Manor—52
89. Marlboro—19
90. Marshall—118
91. Maumee—87
92. Maury—88
93. Mecklenburg—53
94. Meigs—89
95. Melbourne—140
96. Memphis—20
97. Merrimac—67
98. Miami—90
99. Miles—119
100. Miller—120
101. Mohave—141
102. Moody—121
103. Muskinghum—91
104. Myatt—21
105. Nez Perce—142
106. Nipe—161
107. Nacogdoches—22
108. Norfolk—23
109. Ochlockonee—24
110. Oktibbeha—25

111. Olympic—143
112. Ontario—68
113. Ontonagon—92
114. Orangeburg—26
115. Orlando—27
116. Palouse—144
117. Parsons—122
118. Penn—54
119. Phillips—123
120. Pierre—124
121. Poncena—162
122. Pond—145
123. Porters—55
124. Portneuf—146
125. Portsmouth—28
126. Putnam—125
127. Ramona—147
128. Ramsey—56
129. Reeves—148
130. Ritzville—149
131. Rosebud—126
132. Rossmoyne—93
133. Ruston—29
134. San Joaquin—150
135. Sarpy—30
136. Sassafras—31
137. Sharkey—32

138. Shelby—127
139. Sierra—151
140. Soller—163
141. Springdale—152
142. Summit—128
143. Sumter—33
144. Superior—94
145. Susquehana—34
146. Tama—129
147. Tifton—35
148. Tilsit—95
149. Valera—130
150. Victoria—131
151. Volusia—69
152. Watauga—57
153. Waverly—36
154. Webster—132
155. Westmoreland—96
156. Wethersfield—70
157. White Store—58
158. Wickham—59
159. Wilkes—60
160. Willamette—153
161. Williams—133
162. Wooster—71
163. Yukon—155

HINTS FOR THE PRONUNCIATION OF TERMS IN SOIL CLASSIFICATION

The ending of names of orders is regularly *sol* (Latin *solum*, soil); the *o* is pronounced like the *o* in *soluble*; plural forms end in *s* or *z*.

Examples:

1. Entisol (as in b*ent*)
2. Vertisol (as in in*vert*)
3. Incéptisol (as in *incept*ion)

    4. Arídisol (as in *aridi*ty)
    5. Mollisol (as in *molli*fy)
    6. Spodosol (as in r*od*)
    7. Alfisol (as in *Alf*red)
    8. Ultisol (as in *ulti*mate)

The names of the suborders are abstracted from the accented syllables. They are always of two-syllable length, and are accented on the first syllable.

    Examples:

      1. Aquent (as in *aqu*arium)
      2. Psamment (as in *Sam*uel)
      3. Ustent (as in b*ust*)
      4. Udent (as in *you'd*)
      5. Andept (as in *and*)
      6. Umbrept (as in *umbr*ella)
      7. Ochrept (as in *okr*a)
      8. Orthid (as in *orth*opedics)
      9. Argid (as in *Arg*entina)
    10. Rendoll (as in *Rendz*ina)
    11. Alboll (as in *alb*um)
    12. Altoll (as in *alti*tude)
    13. Humod (as in *hum*us)
    14. Plintox (like s*plint*)

The ending for the name of a great group preserves the name of the governing suborder, except for minor euphonic changes:

    a. *ud* loses its initial *y* sound after *r* or *l*:
        Cryudent (cry-you-dent).
        Agrudent (ag-roo-dent).
        Hapludent (hap-loo-dent).
        Vermudoll (verm-you-doll).
        Hapludoll (hap-loo-doll).
        Argudoll (arg-you-doll).
        Agrudalf (ag-roo-dalf).
        Typudalf (type-you-dalf).
        Fragudalf (frag-you-dalf).

Glossudalf (gloss-you-dalf).

Fraglossudalf (frag-loss-you-dalf).

b. *hum* loses its initial h-sound after *h*:

Orthumod (orth-you-mod).

Thermhumod (therm-hue-mod).

The name of a great group is regularly accented on the syllable preceding that part taken from the name of the governing suborder:

Nátrargid, but Nadúrargid.

Glóssudalf, but Fraglóssudalf.

Pronunciation aids for the names of some great groups are given below as examples:

Cryaquent (as in *cry*)

Psammaquent (as if sam-ak-went).

Hydraquent (as in *hyd*rant).

Haplaquent (as in *hap*less).

Quarzopsamment (as in *quartz*).

Agrudent (as in *agr*iculture).

Plaggudent (rhyming with *sag*).

Grumaquent (like *groom*).

Mazaquent (as in *maze*).

Halaquept (as in *hal*lowed).

Fragaquept (as in *frag*ment).

Anthrumbrept (an-thrum-brept as in *an-thro*pology).

Eutrochrept (as if you-tro-crept).

Dystrochrept (as if dis-tro-crept).

Ustochrept (as if us-toe-crept).

Camborthid (as in *Camb*odia).

Durorthid (as in *dur*able).

Calcorthid (as in *Calc*utta).

Salorthid (as in *sal*ary).

Nátrargid (as in *nat*ive).

Nadúrargid (rhyming with en*dure;* first vowel as in *a*dore).

Argalboll (as in *Arg*us).

Vermaltoll (as in *vermi*form).
Humaquod (as in *hum*us).
Ferraquod (as in *ferr*ic).
Placaquod (as if *plaque*).
Thermhumod (as in *thermo*meter).
Typorthod (as in *type*writer).
Albaqualf (as in *alb*um).
Glossaqualf (as in *gloss*ary).
Ochraqualf (as if *okra*).
Umbraqualf (as in *umbr*ella).
Fraglossudalf (as if frag-loss-you-dalf).
Rhodustalf (as in *rhod*odendron).
Ultustalf (as in *ulti*mate).
Plintaquult (as in s*plint*).

ATLANTIC–GULF COASTAL PLAIN

1. CAHABA

Red-Yellow Podzolic – Ultisol – Ochrult – Normochrult – Typic Normochrult.

(Elmore Co., Ala.)

The red, sandy-textured Cahaba series is widely distributed on Coastal Plain terraces from Chesapeake Bay to the Gulf of Mexico. It is a relatively deep, sandy silt loam with good tilth and relatively good drainage. It is widely used for a variety of agricultural crops.

2. COLLINGTON

Regosol – Ultisol – Udult – Hapludult – Typic Hapludult.
(Monmouth Co., N.J.)

The dark brown Collington series is a typical mature soil of eastern New Jersey and northern Maryland. It is derived from glauconite and is relatively high in iron and potassium. It is grayish-brown in color and its profile is well developed. It responds well to fertilizer and is an excellent soil for truck crops.

3. COXVILLE

Low-Humic Gley – Ultisol – Aquult – Ochraquult – Typic Ochraquult.
(App. 2; CP-1) (Craven Co., N.C.)
This dark gray, fine-textured, poorly drained Coxville series is widely distributed over the lower and middle Atlantic–Gulf Coastal Plain. Its poor drainage makes it unattractive for agriculture.

4. CRAWFORD

Brunizem (Prairie) – Vertisol – Ustert – Chromustert – Udic Chromustert.
(Hays Co., Tex.)
The Crawford series occupies a significant area in east central Texas, eastward from the Central granite region and westward from the Black Prairie of Houston soils. It is a black soil with occasional reddish horizons, and is relatively shallow. It is a relatively good soil for cotton, sorghums, and other feed crops.

5. CROWLEY

Low-Humic Gley – Alfisol – Aqualf – Albaqualf – Typic Albaqualf.
(Livingston Par., La.)
This dark gray, light-textured soil series is restricted to the Gulf Coastal Plain. Large areas are found in southern Louisiana and in Arkansas. It has a sandy-textured surface soil underlain by a tight clay subsoil which, when wet, expands and becomes impervious to downward movement of water. This quality makes it attractive for growing rice, since the irrigation waters remain near the surface.

6. DUNBAR

Low-Humic Gley – Ultisol – Ochrult – Normochrult – Aquic Normochrult.

(App. 2; CP-2) (Duplin Co., N.C.)

The Dunbar series is grayish-brown, fine textured and somewhat poorly drained—better drained than the Coxville but not as well drained as the Norfolk. It is widely distributed from the Chesapeake Bay to the Gulf.

7. ELKTON

Low-Humic Gley – Ultisol – Aquult – Ochraquult – Typic Ochraquult.

(Pamlico Co., N.C.)

The Elkton series is found mainly in the Chesapeake Bay region associated with Myatt, Keyport, Leonardtown, and Portsmouth series. It is a light-colored soil with a sandy surface and a much heavier than usual B horizon. Alternating wet and dry seasons account for mottling in the subsoil. This is not an extensive soil and not one of great agricultural value. Much of it remains in hardwoods. Crops, including corn, hay crops, and soy beans, are shallow rooted.

8. GRENADA

Red-Yellow Podzolic – Ultisol – Udult – Paleudult – Rhodic Fragiudalf.

(Grenada Co., Miss.)

The Grenada series is much like the Memphis except that it is derived from thin loëss instead of thick loëss. The chief Grenada belt lies eastward from the Memphis belt on the thinner margins of the loëss area. It is subject to rather severe erosion. It is yellowish in the north, becoming reddish in the south. It is one of the better soils of the area.

9. GREENVILLE

Red-Yellow Podzolic – Ultisol – Udult – Paleudult – Rhodic Paleudult.

(Conecuh Co., Ala.)

The Greenville series often is associated geographically with the Orangeburg. They are quite similar soils except that the Greenville is derived from less sandy material and does not exhibit a gray sandy horizon. The Greenville series responds well to good care and is excellent for a variety of crops.

10. HOUSTON

Grumusol – Vertisol – Aquert – Grumaquert – Typic Grumaquert.

(Dallas Co., Tex.)

The Houston series is derived from limestones and marls. It is high in calcium and has developed under vegetation of tall prairie grass. It has been classed as a Rendzina in the past but more recently as a Grumusol. The Houston is the typical black soil of the Texas Black Prairie and of the Alabama Black Belt. It is an excellent soil for cotton, grain crops, and pasture.

11. HYDE

Humic-Gley – Inceptisol – Aquept – Umbraquept – Typic Umbraquept.

(App. 2; CP-3) (Hyde Co., N.C.)

The Hyde series is not widely distributed. It is found in the wettest part of the outer Coastal Plain of Virginia and North Carolina. It is a very poorly drained, jet black soil with high organic matter content. Hyde loam grades imperceptibly into Hyde muck.

12. KATY

Brunizem (Prairie) – Alfisol – Udalf – Pauledalf – Albaquie Pauledalf.

(Hardin Co., Tex.)

The Katy series occupies a belt immediately north of the Lake Charles soil belt in eastern Texas and southern Louisiana. These are light-colored soils with a silty A horizon and a heavy, hard, tough clay B horizon. The Katy series is somewhat transitional between the black Rendzina-like soils and the light-colored, sandy soils of the Gulf Coast area.

13. LAKE CHARLES

Grumusol – Vertisol – Aquert – Grumaquert – Typic Grumaquert.

(Jefferson Davis Par., La.)

The Lake Charles series occupies a coastal area of southwestern Louisiana and southeastern Texas. The soil is derived from marls and marly materials, similar to the parent materials of the Houston series. It is a dark-colored, imperfectly drained grassland soil formerly classed as Rendzina. The Lake Charles series is an excellent pasture soil. Much of it is used for production of rice, cotton, and sorghum.

14. LAKELAND

Regosol – Entisol – Psamment – Quarzipsamment – Typic Quarzipsamment.

(Harnett Co., N.C.)

The Lakeland series is a somewhat excessively drained, sandy soil of the inner Coastal Plain from Virginia southward. It has a loose sandy surface layer and a yellowish-brown subsoil. It is a dominant soil in several areas: Carolina Sand Hills, parts of Florida, and parts of eastern Texas. It formerly was called "Norfolk Sand."

15. LEON

Ground-Water Podzol – Spodosol – Aquod – Humaquod – Typic Humaquod.

(App. 2; CP-4) (Clinch Co., Ga.)

The Leon series is the most widely distributed hardpan (organic) soil of the Atlantic–Gulf Coastal Plain. It is light-colored soil exhibiting a "pepper-and-salt" appearance in cultivated fields. It has very poor agricultural quality. Blueberries do well on it.

16. LEONARDTOWN

Ground-Water Podzol – Ultisol – Aqult – Fragiaqult – Typic Fragiaqult.

(District of Columbia)

The Leonardtown series is developed from ancient crystalline rock on flat upland areas. It is poorly drained, imperfectly developed, and often develops a hardpan. It is associated geographically with Chester and Manor soils of the Virginia–Maryland Piedmont.

17. LYNCHBURG

Low-Humic Gley – Ultisol – Ochrult – Normochrult – Aquic Normochrult.

(App. 2; CP-5) (Pender Co., N.C.)

The Lynchburg series is relatively poorly drained. However, with proper artificial drainage it is a good agricultural soil. It is a deep gray to pale yellow, light-textured soil found on the Coastal Plain from North Carolina to Mississippi.

18. MAGNOLIA

Red-Yellow Podzolic – Ultisol – Ochrult – Normochrult – Rhodic Normochrult.

(App. 2; CP-6) (Wayne Co., N.C.)

The grayish-brown to reddish-brown Magnolia soil is one of the better soils of the Atlantic Coastal Plain. It is well drained and light textured. Its acreage is not large.

19. MARLBORO

Red-Yellow Podzolic – Ultisol – Udult – Paleudult – Typic Paleudult.

(App. 2; CP-7) (Wilson Co., N.C.)

The deep gray to brownish-gray Marlboro sandy loam is one of the Coastal Plain's best soils. It is excellent for tobacco, cotton, and peanuts.

20. MEMPHIS

Red-Yellow Podzolic – Alfisol – Udalf – Normudalf – Typic Normudalf.

(Shelby Co., Tenn.)

The light-colored, light-textured Memphis series is dominant over wide areas of thick loess from western Kentucky to Louisiana. It is yellowish to reddish and is subject to rather severe erosion. Memphis silt loam is one of the best soils in western Tennessee and northern Mississippi. Much of it is used for growing cotton, corn, soy beans, and other grain crops.

21. MYATT

Low-Humic Gley – Ultisol – Aquult – Ochraquult – Typic Ochraquult.

(Princess Anne Co., Va.)

The Myatt series is one of several terrace soils found on the Coastal Plains southward from Chesapeake Bay. It is poorly drained, sandy to sandy loam soil. The surface soil is gray to light gray, and the subsoil is distinctly mottled. It is associated with low river terraces and sometimes occupies low marine terraces. It is suitable for shallow-rooted crops only. Its area is not extensive.

22. NACOGDOCHES

Red-Yellow Podzolic – Ultisol – Ochrult – Rhodochrult – Typic Rhodochrult.

(Nacogdoches Co., Tex.)

Nacogdoches soils occupy important areas in the Gulf Coastal Plain of Louisiana and eastern Texas. They resemble somewhat the Collington soils of New Jersey and Maryland in that they are derived from Tertiary sands containing glauconite. The Nacogdoches soil, predominantly a fine sandy loam, is much more aged than the Collington soil, and is one of the most distinctly red soils in the United States. It handles well and is relatively good, agriculturally.

23. NORFOLK

Red-Yellow Podzolic – Ultisol – Ochrult – Normochrult – Typic Normochrult.

(App. 2; CP-8) (Johnston Co., N.C.)

The deep, well-drained, pale yellow Norfolk series is the classic example of Yellow Podzolic soils of the Coastal Plain. It is found widely from Virginia to Texas. It is one of the best soils of the region. More tobacco and peanuts are produced on Norfolk soil than on any other soil in the South.

24. OCHLOCKONEE

Alluvial – Entisol – Orthent – Haplorthent – Typic Haplorthent.

(Washington Co., Ala.)

The Ochlockonee series of alluvial soil is found widely on floodplains of small and medium-sized rivers entering the Gulf. It is gray to brownish in color and poorly drained. Corn and other shallow-rooted crops do well on this soil when they escape flooding.

25. OKTIBBEHA

Red-Yellow Podzolic – Vertisol – Aquert – Mazaquert – Ochrultic Mazaquert.

(Lowndes Co., Ala.)

The Oktibbeha is associated with Houston and Sumter series. It is found particularly on the southern margin of the Alabama Black Belt where the marl approaches the surface and is covered with a thin layer of younger sediments. Thus the A horizon of the Oktibbeha series is derived from the overlying sediments while the lower horizons are derived from marl. It has a gray or light brown surface and a brown or reddish subsoil. It is widely used for grains, feed crops, and pasture. Much of it has suffered from severe erosion under cotton culture.

26. ORANGEBURG

Red-Yellow Podzolic – Ultisol – Ochrult – Normochrult – Rhodic Normochrult.

(Wilkinson Co., Ga.)

The red Orangeburg series is widespread over the inner Coastal Plain of South Carolina and Georgia. It also is found in Mississippi, Louisiana, and Arkansas but is less typical of these areas. It is derived from impure Tertiary limestones and normally is a sandy loam. It resembles the Norfolk except that the Orangeburg is distinctly red and its B horizon is heavier than that of the Norfolk. The Orangeburg series is one of the best Coastal Plain soils. It is excellent for cotton and other field crops and for peach orchards.

27. ORLANDO

Humic-Gley – Inceptisol – Umbrept – Haplumbrept – Quartzipsammentic Haplumbrept.

(Orange Co., Fla.)

The Orlando series is found in the interior of central Florida. It is a dark-colored soil developed under grass vegetation. It is light-textured, with sandy loams and silt loams predominant. The Orlando series is one of Florida's best soils. Much of the truck farming of Florida (for example, that of the Orlando–Sanford area) utilizes this excellent soil.

28. PORTSMOUTH

Humic-Gley – Inceptisol – Aquept – Umbraquept – Typic Umbraquept.

(App. 2; CP-9) (Tyrrell Co., N.C.)

The very poorly drained Portsmouth soil is dark gray to black and is usually sandy to loamy in texture. Often it borders marshes and swamps on the outer Coastal Plain.

29. RUSTON

Red-Yellow Podzolic – Ultisol – Ochrult – Normochrult – Typic Normochrult.

(App. 2; CP-10) (Johnston Co., N.C.)

The reddish-yellow to reddish-brown Ruston soil is sometimes referred to as the "red Norfolk." It is deep, well drained, and excellent for a variety of crops, including tobacco. It is often associated with Norfolk, usually on steeper slopes.

30. SARPY

Alluvial – Entisol – Psamment – Normipsamment – Cumulie Normipsamment.

(Tensas Par., La.)

The Sarpy series occupies the natural levees along the Mississippi River and its major tributaries. It has a sandy texture and is relatively well drained. It is yellowish at the surface but becomes lighter in color with depth. The parent material is relatively high in calcium and potassium, and so the soil is quite productive when handled properly. A large proportion is used for crops.

31. SASSAFRAS

Red-Yellow Podzolic – Ultisol – Udult – Hapludult – Typic Hapludult.

(Queen Annes Co., Md.)

The Sassafras series is the dominant Gray-Brown Podzolic soil of the unglaciated portion of the Coastal Plain northward from Chesapeake Bay. It has a well-developed sandy profile ranging in color from yellowish-gray at the surface to reddish-brown in the subsoil. It has good tilth, responds well to fertilizer, and is suited to truck crops as well as to common field crops.

32. SHARKEY

Alluvial – Vertisol – Aquert – Grumaquert – Entic Grumaquert.

(Washington Co., Miss.)

The Sharkey series is the dominant soil of the back swamps along the Mississippi. It is an immature, poorly drained, dark clay loam or clay. It is relatively fertile but hard to manage. Flooding is a major problem.

33. SUMTER

Grumusol – Inceptisol – Ochrept – Entrochrept – Rendallic Entrochrept.

(Sumter Co., Ala.)

The Sumter series is much like the Houston except that it is a much shallower soil. In some instances erosion has exposed the marl at the surface. The less eroded Sumter soils are excellent for a variety of crops and for pasture.

34. SUSQUEHANA

Regosol – Inceptisol – Ochrept – Dystrochrept – Aquic Dystrochrept.

(Choctaw Co., Ala.)

The Susquehana series is found widely on the Coastal Plain from Maryland to Texas. Its greatest areas of concentration are in the Gulf rather than the Atlantic portion of the Coastal Plain. It is a rather thin soil with a sandy A horizon and a clay C horizon. The B horizon is missing. It is difficult to handle and is not a highly important agricultural soil.

35. TIFTON

Ground-Water Lateritic – Ultisol – Ochrult – Normochrult – Plintic Normochrult.

(Tift Co., Ga.)

The Tifton series is associated geographically with Norfolk, Orangeburg, and Greenville soils. Its area of greatest concentration is the inner Coastal Plain of South Carolina and Georgia. The Tifton series differs from its associates in that small red iron concretions and other ironstone fragments strewn over its surface give it a pimply appearance. The Tifton soil, when not too stony, is a relatively good general-purpose soil.

36. WAVERLY

Alluvial – Inceptisol – Aquept – Ochraquept – Cumulic Ochraquept.

(Richland Par., La.)

The Waverly series is the major soil on the wide floodplains of the Gulf Coast rivers. The major areas are associated with the Mississippi, the Alabama, and the Arkansas Rivers. It is a poorly drained soil with only faintly developed profile features. It is relatively high in organic matter and is light in color. Flooding is a major handicap in the use of Waverly soil.

SOUTHERN APPALACHIANS: BLUE RIDGE, PIEDMONT, AND GREAT APPALACHIAN VALLEY

37. ALAMANCE

Red-Yellow Podzolic – Ultisol – Ochrult – Normochrult – Typic Normochrult.

(Alamance Co., N.C.)

This yellow, moderately deep soil is derived from the Carolina Slates of the Piedmont. The low angle of slope is usually less than 5°. The surface soil is a gray to yellow, friable, smooth silt loam, occasionally slightly sandy or gravelly. The somewhat heavier-textured subsoil is smooth when moist, "slick" or sticky when wet, and hard when dry. The Alamance series is widely distributed throughout the Slate Belt and is moderately good for a variety of crops, for pasture, and for timber.

38. APPLING

Red-Yellow Podzolic – Ultisol – Ochrult – Normochrult – Typic Normochrult.

(App. 2; P-1) (Wake Co., N.C.)

This yellowish-red soil, derived from acid crystalline rocks, is widespread over the Piedmont from Virginia to Alabama.

39. ASHE

Gray-Brown Podzolic – Inceptisol – Ochrept – Dystrochrept – Typic Dystrochrept.

(App. 2; M-1) (Ashe Co., N.C.)

This shallow, stony, yellowish-brown series is found at high elevations in the southern Appalachians.

40. AVERY

Podzol – Inceptisol – Ochrept – Dystrochrept – Typic Dystrochrept.

(App. 2; M-2) (Avery Co., N.C.)

The Podzol is found at high elevations in the southern Blue Ridge region. It is brownish, moderately deep, and relatively stony. The total acreage is small.

41. CECIL

Red-Yellow Podzolic – Ultisol – Ochrult – Normochrult – Typic Normochrult.

(App. 2; P-2) (Patrick Co., Va.)

This is one of the most widely distributed soils of the Piedmont, extending from the Potomac River to Alabama. It is the classic, red, upland soil of the region.

42. CHESTER

Gray-Brown Podzolic – Ultisol – Udult – Hapludult – Typic Hapludult.

(York Co., Pa.)

The Chester series is a dominant Gray-Brown Podzolic soil of the upper Piedmont and lower reaches of the southern Blue Ridge. It is a maturely developed soil from ancient crystalline rocks. It is better developed in the Virginia section than further south, becoming reddish-brown in North Carolina. It occupies a rolling relief and is fair to good for general farming.

43. CLARKSVILLE

Gray-Brown Podzolic – Ultisol – Ochrult – Normochrult – Typic Normochrult.

(Knox Co., Tenn.)

The Clarksville series is classed as a Gray-Brown Podzolic soil although it is transitional toward the Red-Yellow Podzolic group. It is yellowish to reddish, with texture ranging from sand to clay loam. It is derived from cherty limestones and in some instances exhibits a high degree of stoniness. It occupies narrow valleys and steep ridges in the Great Appalachian Valley and on the Highland Rim of Tennessee and Kentucky. It is only fair for crops.

44. CONGAREE

Alluvial – Entisol – Orthent – Haplorthent – Typic Haplorthent.
(App. 2; P-3) (Chatham Co., N.C.)
The Congaree soil is widely distributed over the Piedmont, occupying narrow strips of alluvium along most streams. It is brownish and is poorly drained. Shallow-rooted crops do well on it when flooding can be avoided.

45. DAVIDSON

Red-Yellow Podzolic – Ultisol – Ochrult – Rhodochrult – Typic Rhodochrult.
(App. 2; P-4) (Cabarrus Co., N.C.)
The Davidson series is derived from dark-colored basic rocks. It is widely distributed over the Piedmont but usually occurs in relatively small areas. Its deep red color, relatively fine texture, and good tilth make it an exceptionally good agricultural soil.

46. DECATUR

Red-Yellow Podzolic – Ultisol – Ochrult – Rhodochrult – Typic Rhodochrult.
(Grainger Co., Tenn.)
The Decatur series is a Tennessee Valley soil derived from chert-free limestones. Occasionally it has a yellowish surface but usually it is a strong red throughout. It is a red member of the Red-Yellow Podzolic group, well matured and corresponding somewhat to the Gray-Brown Podzolic Hagerstown series

further north in the Great Valley. It is one of the better general farming soils of the Tennessee Valley.

47. DURHAM

Red-Yellow Podzolic – Ultisol – Ochrult – Normochrult – Typic Normochrult.

(App. 2; P-5) (Durham Co., N.C.)

The Durham series is the yellow member of Piedmont soils derived from acid crystalline rocks. It is widely distributed over the Piedmont from Virginia to Georgia. Its fine sandy loam texture makes it suitable for relatively wide agricultural use.

48. GEORGEVILLE

Red-Yellow Podzolic – Ultisol – Ochrult – Normochrult – Typic Normochrult.

(App. 2; P-6) (Orange Co., N.C.)

The bright red Georgeville series resembles the Cecil in color and in texture. It is the characteristic, mature, upland soil of the Carolina Slates.

49. HAGERSTOWN

Gray-Brown Podzolic – Ultisol – Ochrult – Normochrult – Umbric Normochrult.

(Washington Co., Md.)

The brown to reddish-brown Hagerstown loam is one of the superior general farming soils in eastern United States. It is characteristic of limestone valleys and basins and is found widely in (a) the Great Appalachian Valley from Tennessee to Pennsylvania, (b) the Piedmont of Pennsylvania and Maryland, (c) the Nashville Basin, and (d) limestone areas of Kentucky and Indiana. It is derived from impure limestone and it ranges from shallow to relatively deep. It usually occupies rolling relief and is well suited to general farming and pasturing.

50. IREDELL

Planosol – Alfisol – Udalf – Normudalf – Albaqualfic Normudalf.

(App. 2; P-7) (Guilford Co., N.C.)

The Iredell series, like the Davidson and Mecklenburg series, is derived from dark-colored, basic rocks. It is the least developed of the three and is inferior to the other two for agricultural purposes. It is usually found on moderately steep slopes.

51. MADISON

Red-Yellow Podzolic – Ultisol – Ochrult – Normochrult – Typic Normochrult.

(App. 2; P-8) (Surry Co., N.C.)

The grayish to reddish-brown, moderately deep, clay loam Madison series is derived from micaceous granitic rock. It is widely distributed over the Piedmont, although its total acreage is not great.

52. MANOR

Gray-Brown Podzolic – Inceptisol – Ochrept – Dystrochrept – Typic Dystrochrept.

(Lancaster Co., Pa.)

The brown to reddish-brown Manor series is much like the Chester except that it occupies relatively steep slopes and is not maturely developed because of loss of surface material by excessive erosion. It is relatively fertile but requires erosion control. It is found in the Chesapeake Bay area of Maryland and Virginia.

53. MECKLENBURG

Red-Yellow Podzolic – Alfisol – Udalf – Normudalf – Ochrultic Normudalf.

(App. 2; P-9) (Mecklenburg Co., N.C.)

The dark gray to reddish-brown, moderately deep Mecklenburg loam is intermediate in development of the soils derived from dark-colored basic rocks. It is an excellent Piedmont soil.

54. PENN

Lithosol – Alfisol – Udalf – Hapludalf – Ultic Hapludalf.

(Bucks Co., Pa.)

The Penn series is the typical Triassic lowland soil found northward from Virginia. In New Jersey it is derived from red shales and gets its red color from the parent rock and not from oxidation. It also is found in colors ranging from gray to brown. It is relatively shallow soil and often has a sandy texture. In the glaciated areas of Triassic sediments, the corresponding soil is mapped as Wethersfield and other similar series.

55. PORTERS

Gray-Brown Podzolic – Ultisol – Ochrult – Normochrult – Udalfic Normochrult.

(App. 2; M-3) (Transylvania Co., N.C.)

The Porters series is one of the more widely distributed Gray-Brown Podzolic soils of the southern Blue Ridge Mountains. It is shallow to moderately deep and is frequently quite stony.

56. RAMSEY

Lithosol – Inceptisol – Ochrept – Dystrochrept – Lithic Dystrochrept.

(App. 2; M-4) (Cherokee Co., N.C.)

The shallow brown Ramsey series is widely distributed in the Blue Ridge. It is derived from fine-grain rocks, including shales, slates, and quartzites. It is moderately stony and usually occupies steep slopes. It is not important agriculturally.

57. WATAUGA

Gray-Brown Podzolic – Ultisol – Ochrult – Normochrult – Typic Normochrult.

(App. 2; M-5) (Watauga Co., N.C.)

The Watauga series is limited in acreage in the Blue Ridge Province. It occupies relatively smooth areas in the lower mountain slopes. It is a moderately deep, grayish-brown loam derived from micaceous rocks.

58. WHITE STORE

Planosol – Ultisol – Ochrult – Normochrult – Aquic Normochrult.

(App. 2; P-10) (Lee Co., N.C.)

The White Store series is the characteristic mature soil of the Triassic basins of the Virginia-Carolina Piedmont. It is a moderately deep, gray to pale yellow, friable, silt to sandy loam.

59. WICKHAM

Red-Yellow Podzolic – Ultisol – Ochrult – Normochrult – Typic Normochrult.

(App. 2; P-11) (Richmond Co., N.C.)

The deep, brown, sandy loam Wickham series is distributed widely on river terraces of the Piedmont.

60. WILKES

Lithosol – Inceptisol – Ochrept – Dystrochrept – Ruptic Alfic Dystrochrept.

(App. 2; P-12) (Caswell Co., N.C.)

The gray to brownish-yellow Wilkes series is widely distributed over the Piedmont from Virginia to Alabama. It is found on steep slopes and is derived from a mixture of acid and basic crystalline rocks.

NORTHEASTERN UNITED STATES

61. CARIBOU

Podzol – Spodosol – Orthod – Háplorthod – Alfic Haplorthod.

(Coos Co., N.H.)

The Caribou is New England's best Podzol. It is found in

northern Maine, northern New Hampshire, and in restricted areas in other parts of the region. It is derived from calcareous glacial till and is developed under heavy forest growth, largely hardwoods—particularly birch and maple. It has a maturely developed profile. It ranges widely in depth and stoniness. The Caribou series comprises much of the prize potato land of Aroostook Co., Maine.

62. DEKALB

Lithosol – Inceptisol – Ochrept – Dystrochrept – Typic Dystrochrept.

(Delaware Co., N.Y.)

The Dekalb series differs from other soils associated with it geographically in that it is derived only in part from glacially deposited materials. Whether ice-deposited or not, the parent materials of the Dekalb consist of sandstones and shales. This is not an important agricultural soil and its total extent is not great. One of its chief areas of occurrence is on the Allegheny Plateau of south central New York and northern Pennsylvania.

63. DUNKIRK

Ground-Water Podzol – Alfisol – Udalf – Hapludalf – Glossoboric Hapludalf.

(Erie Co., N.Y.)

The Dunkirk series is found mainly along the south shore of Lake Ontario and to some extent along the south shore of Lake Erie. It is a young, imperfectly developed soil derived from water-laid materials. It is brownish in color and has a heavy texture tending toward hardpan. It is not a good agricultural soil. Crops include some hay crops, buckwheat, and other grains.

64. GLOUCESTER

Brown Podzolic – Spodosol – Orthod – Haplorthod – Entic Haplorthod.

(Essex Co., Mass.)

The Gloucester series is the typical, mature Brown Podzolic soil of the southern New England upland. It is derived principally from glacially deposited materials from crystalline schists and gneisses. Weathering of these rocks is shallow and the Gloucester profile is incompletely developed. Profile features vary widely within the series, horizons being very distinct in some and only faint in others. The extent of its development increases to the south. The soil is brown throughout, becoming deeper brown with depth. Sharp horizon transitions are noticed in the New Jersey section. The Gloucester series is widely used agriculturally.

65. HERMON

Podzol – Spodosol – Orthod – Haplorthod – Typic Haplorthod.

(Franklin Co., N.Y.)

The Hermon series is second to Caribou among New England's Podzols. Its profile is less developed than that of the Caribou. In general, it occupies relatively rugged terrain, is quite stony, and is greatly inferior to the Caribou for agricultural crops.

66. LORDSTOWN

Gray-Brown Podzolic – Inceptisol – Ochrept – Dystrochrept – Typic Dystrochrept.

(Susquehanna Co., Pa.)

The Gray-Brown Podzolic Lordstown series is found in northern Pennsylvania and southern New York in the glaciated part of the Appalachian Plateau. It is derived from glacially laid sandstone and shale materials. In many places it occupies relatively steep slopes. Its profile is not maturely developed and it is a rather inferior agricultural soil. In many respects the Lordstown series resembles the Muskinghum series.

67. MERRIMAC

Brown Podzolic – Spodosol – Orthod – Haplorthod – Entic Haplorthod.

(Kent Co., R.I.)

The Merrimac series is the dominant New England soil developed on glacial outwash and to some extent on river terraces. It is an excessively drained soil comprised of sand and gravelly materials. It is a yellowish to distinctly brown soil with good depth and tilth. Much of it is used for truck crops.

68. ONTARIO

Gray-Brown Podzolic – Alfisol – Udalf – Hapludalf – Glossoboric Hapludalf.

(Onondaga Co., N.Y.)

The Ontario series is widely distributed in the low lying areas north of the Allegheny Plateau in Pennsylvania and New York. In this general area limestones are dominant and the glacial till is calcareous. There is much water-laid material, but the Ontario series is derived from calcareous ice-deposited material. In general the profile is maturely developed; in some places drainage is poor and the profile lacks maturity. It is a heavy-textured brown soil with considerable organic material. It is used extensively for vegetables, buckwheat, and hay crops.

69. VOLUSIA

Ground-Water Podzol – Inceptisol – Aquept – Fragiaquept – Aeric Fragiaquept.

(Erie Co., Pa.)

The Volusia series is an immature soil in western New York and Pennsylvania. It is developed from glacially deposited sandstone and shale materials under excessive moisture but not high water-table conditions. It tends toward hardpan development. It is a light-colored soil and is similar to the Lordstown and Mahoning series.

70. WETHERSFIELD

Podzol – Inceptisol – Ochrept – Fragiochrept – Typic Fragiochrept.

(Hampden Co., Mass.)

The Wethersfield series is found in glaciated areas of Triassic lowlands of New York and New England. It is common in the Connecticut lowlands. It sometimes is considered the counterpart of the Penn series in the unglaciated Triassic areas found further south. The Wethersfield series usually occurs on the flat to rolling till-plain where local Triassic sandstones and shales supply the material. It also is found on drumlins of similar areas. It is a relatively light-textured soil, and it is a fairly good soil for grains, hay, and pasture.

71. WOOSTER

Gray-Brown Podzolic – Alfisol – Udalf – Fragiudalf – Typic Fragiudalf.

(Mercer Co., Pa.)

The Wooster series is widely distributed in northern Ohio and adjacent areas. It is a mature soil derived from glacially deposited sandstone and calcareous shale materials. It developed under excessive moisture conditions but it is a well-drained soil with mature profile features. It is a deep, brown soil with sandy loam to clay loam texture, and is one of the best agricultural soils of the Appalachian Plateau area.

EAST CENTRAL UNITED STATES: APPALACHIAN PLATEAU TO MISSISSIPPI RIVER

72. BAXTER

Gray-Brown Podzolic – Ultisol – Udult – Paleudult – Typic Paleudult.

(Stewart Co., Tenn.)

The Baxter series is widely distributed in two geographical areas: (1) Tennessee–Kentucky area westward from the limestone basins and (2) Ozark Ouachita area. It is a light, yellowish-brown soil derived from cherty limestones and from sandstones. Its southern location gives it a reddish rather than brownish color. The Baxter soil is immaturely developed. It is subject to severe erosion, is hard to cultivate, and is not a highly prized agricultural soil.

73. BELLEFONTAINE

Gray-Brown Podzolic – (Not reactivated into the 7th Approximation system).

(Henry Co., Ohio)

The Bellefontaine series is maturely developed in the rolling glacial country of western Ohio. It is derived from calcareous Wisconsin till of Niagara limestone origin. The well-drained Bellefontaine silt loam is an excellent soil for a variety of grain and feed crops.

74. BELTRAMI

Gray Wooded – Alfisol – Boralf – Entroboralf – Typic Entroboralf.

(Hubbard Co., Minn.)

The Beltrami series is one of the best developed Gray Wooded soils. It is widely distributed in northern Minnesota and northward into Canada. It is one of the most maturely developed soils in this part of the United States, acquiring a bleicherde layer approaching a foot in thickness in places. The Beltrami series is derived from highly calcareous glacial till. It is one of the best soils of the area for agricultural use.

75. BROOKSTON

Half-Bog – Mollisol – Aquoll – Argiaquoll – Typic Argiaquoll.

(Fulton Co., Ind.)

The Brookston series is much like the Clyde series, except that Brookston is grayish-brown and is somewhat better drained than the Clyde. They are associated geographically with Crosby and Miami soils.

76. CINCINNATI

Gray-Brown Podzolic – Alfisol – Udalf – Fragiudalf – Aeric Fragiudalf.

(Martin Co., Ind.)

The Cincinnati series is the typical maturely developed Gray-Brown Podzolic soil in southern Ohio and Indiana. It is found in the area of Illinoian glaciation, where a relatively thin layer of loëss covers the glacial till. It occupies gently sloping terrain, is well drained, and is a relatively good agricultural soil. It has a light-textured A horizon with a brown to reddish-brown B horizon.

77. CLERMONT

Planosol – Alfisol – Aqualf – Fragiaqualf – Typic Fragiaqualf.
(Clermont Co., Ohio)
The immature Clermont soil is found in the poorly drained areas of Illinoian till. In general it is located north of the Cincinnati series, where the loëss cover is somewhat thicker. The A and B horizons are indistinct. The soil has limited agricultural value.

78. CLYDE

Humic-Gley – Mollisol – Aquoll – Haplaquoll – Typic Haplaquoll.
(Wood Co., Ohio)
The Clyde series is also associated with the Miami series, both being derived from calcareous Wisconsin till. The Clyde soil is poorly drained naturally, but much of it has been drained artificially (since the 1880's). The drained Clyde series is a black soil with poorly developed profile features. It is fairly productive agriculturally.

79. CROSBY

Gray-Brown Podzolic – Alfisol – Aqualf – Ochraqualf – Aeric Ochraqualf.
(Randolph Co., Ind.)
The Crosby series is much like the Miami and is associated with it geographically. The chief difference is that the Crosby is lighter in color and has a lower B horizon that is poorly drained and much heavier than that of the Miami, tending

toward hardpan development. The Crosby series is inferior to the Miami series for agricultural use.

80. FAIRMONT

Rendzina – Mollisol – Udoll – Hapludoll – Typic Hapludoll.

(Adams Co., Ohio)

Fairmont soil occupies steep slopes below the outcrops of Illinoian glacial till. It is derived from disintegrated bedrock, largely limestones and calcareous shales. The Fairmont is a very thin soil with scarcely any profile features. It is agriculturally unimportant.

81. FOX

Gray-Brown Podzolic – Alfisol – Udalf – Hapludalf – Typic Hapludalf.

(Newaygo Co., Mich.)

The Fox series is an excessively drained light-colored soil derived from gravelly glacial outwash largely of limestone origin. It is maturely developed. The A and B horizons are much like the Miami but the C horizon is calcareous gravel. It is found most extensively in southern Michigan. Much of it is in timber.

82. HANCEVILLE

Lithosol – Ultisol – Udult – Rhodudult – Typic Rhodudult.

(Van Buren Co., Ark.)

The chief area of Hanceville soils is the Ozark-Ouachita upland of Arkansas. Small areas are also found in the Appalachian Plateau of Alabama, Tennessee, and Kentucky. This is a grayish-brown to reddish soil derived from sandstones and shales. It is a well-drained soil with good tilth. It is low in organic matter and is only fair for crops.

83. HARTSELLS

Lithosol – Ultisol – Udult – Hapludult – Typic Hapludult.

(Cumberland Co., Tenn.)

The yellowish Hartsells series is a poorly drained, upland soil derived from sandstones and shales. It is widely distributed on the Appalachian Plateau from Alabama to Kentucky and to a limited extent in the Ozark upland. Hartsells and Hanceville soils are often associated geographically, owing their distinction largely to drainage. The poorly drained, imperfectly developed, sandy Hartsells soil is one of the poorest agricultural soils of the Appalachian Plateau region.

84. HUNTINGTON

Alluvial – Mollisol – Udoll – Hapludoll – Fluventic Hapludoll.

(Marshall Co., Ky.)

Huntington is a widely distributed floodplain soil in many river valleys of the Middle West. The major areas are associated with major rivers, of which the Ohio and its tributaries provide a good example. It has a better developed profile than usual for floodplain soils. Its A horizon is a brownish silty or sandy loam; its B horizon is somewhat heavier textured and a lighter brown in color. Mottling is common. The Huntington series is highly productive and very important agriculturally. It is highly susceptible to flooding.

85. LOWELL

Gray-Brown Podzolic – Alfisol – Udalf – Hapludalf – Typic Hapludalf.

(Shelby Co., Ky.)

The Lowell series is dominant in the Kentucky area surrounding the Blue Grass, for the most part west of the Muskinghum area. It is derived from argilaceous chert-free limestone. It is a light-colored soil, becoming faintly red in its southern reaches. It is imperfectly developed and the B horizon is often missing. The C horizon is tough, sticky, and plastic. The Lowell soil usually develops where the mantle is thin (less than 4'). Where the mantle is thick, the Lowell profile resembles that of the Hagerstown, except Hagerstown has a friable C horizon. The Lowell soil is only fair for agricultural crops.

86. MAHONING

Humic-Gley – Alfisol – Aqualf – Ochraqualf – Aeric Ochraqualf.

(Allegany Co., N.Y.)

The Mahoning series is an immature soil developed under excessive moisture conditions (but not a high water table). It is derived from sandstones and calcareous shales. The chief area in which it occurs is in northeastern Ohio. It is not a highly important agricultural soil.

87. MAUMEE

Half-Bog – Mollisol – Aquoll – Haplaquoll – Typic Haplaquoll.

(Kankakee Co., Ill.)

The Maumee series is a Half-Bog soil, distributed widely through the Great Lakes area, particularly in northern Indiana and Illinois. It is associated with marsh and swamp land and in some places grades into muck. The Maumee series is very dark colored, containing a large amount of organic matter. Drainage in general is poor. In one area of Maumee soil in Illinois, underlying gravel provides good drainage and a brown silt loam has developed. Much artificial drainage has been provided, particularly in Indiana. Crops consist principally of corn, small grains, and soy beans.

88. MAURY

Gray-Brown Podzolic – Ultisol – Udult – Paleudult – Humic Paleudult.

(Fayette Co., Ky.)

The Maury series is found principally in the Nashville and the Blue Grass Basins and is derived from Lower Ordovician limestone. It resembles the Hagerstown in some respects except that it has a much higher content of available phosphorus. The rich, brown Maury clay loam is one of the world's most productive upland soils. It occupies gently rolling terrain and is well drained. It is famous for blue grass pastures and racehorse farms. It is an excellent soil for Burley tobacco, corn, and other crops.

89. MEIGS

Gray-Brown Podzolic – (Not reactivated into 7th Approximation system).

(Washington Co., Ohio)

The dominant area of Meigs soil is on the Allegheny Plateau of Pennsylvania, Ohio, and West Virginia, southward from the area of Westmoreland soils. The two are much alike except that the Meigs is derived from shales and is inferior to the Westmoreland.

90. MIAMI

Gray-Brown Podzolic – Alfisol – Udalf – Hapludalf – Typic Hapludalf.

(Noble Co., Ind.)

The Miami series and its associates comprise one of the most important groups of soils in the Middle West. They are dominant soils in western Ohio, central northern Indiana, northeastern Illinois, southern Michigan, and southeastern Wisconsin. Miami soil is derived from calcareous till of Wisconsin age. It occupies level to rolling terrain, is well drained, and has maturely developed profile features. It is brown from the surface downward and is only mildly acid. It is an excellent agricultural soil, being suited to a variety of crops.

91. MUSKINGHUM

Lithosol – Inceptisol – Ochrept – Dystrochrept – Typic Dystrochrept.

(Clarion Co., Pa.)

The Muskinghum series is widely distributed over the Appalachian Plateau and the upper reaches of the Ohio Valley. It occupies very uneven relief, and in most places is severely eroded. It is derived from sandstones and shales, is young and imperfectly drained, and has a poorly developed profile. It is yellowish rather than deep brown and is poor for crops. Much of it is in timber.

92. ONTONAGON

Gray Wooded – Alfisol – Boralf – Entroboralf – Typic Entroboralf.

(Cheboygan Co., Mich.)

The Ontonagon series is a prominent Gray Wooded soil of northern Michigan, bordering the three upper Great Lakes. It is maturely developed in many places but podzolization is not as advanced as in the Beltrami series. It is quite sandy in many places and grades into unclassified sands. It is not a strong agricultural soil, and much of it is timbered or unused.

93. ROSSMOYNE

Gray-Brown Podzolic – Alfisol – Udalf – Fragiudalf – Aquic Fragiudalf.

(Gibson Co., Ind.)

The Rossmoyne series is associated with Clermont and Cincinnati soils. It is found in southern Ohio and Indiana near the southern boundary of glaciation where Illinoian till has a relatively thin covering of loess. It is better drained than Clermont and not as well drained as Cincinnati. The A horizon is well drained, the B horizon is poorly drained and has some mottling. It appears to have developed as the result of headward erosion improving the drainage of Clermont soil.

94. SUPERIOR

Podzol – Spodosol – Orthod – Haplorthod – Alfic Haplorthod.

(Kewaunee Co., Wis.)

The Superior series is found mainly in southern Michigan and eastern Wisconsin. It is derived from calcareous till and occupies flat to gently rolling morainic areas. It has a submaturely developed profile and is poorly drained except where terrain is rolling. Normally it has a heavy texture, but some areas are sandy.

95. TILSIT

Red-Yellow Podzolic – Ultisol – Udult – Fragiudult – Typic Fragiudult.

(Dubois Co., Ind.)

The Tilsit series is widespread from the Appalachian Plateau into the Midwestern States. A sizable area is found in southern Indiana and northern Kentucky. The soil is derived from decomposed sandstones and shales on flat relief. It is similar to the Clermont series but is somewhat better drained. Tilsit soils' have developed under timber cover and have a normally developed brownish A horizon. The B horizon, however, shows poorer drainage and is mottled, becoming gray.

96. WESTMORELAND

Gray-Brown Podzolic – Alfisol – Udalf – Hapludalf – Ultic Hapludalf.

(Washington Co., Pa.)

The Westmoreland series is found principally in the Allegheny Plateau in Pennsylvania, Ohio, and West Virginia. It is almost completely surrounded by the Muskinghum soil and is much like it except that Westmoreland is derived from limestone. Its profile is intermediate to poorly developed. The better soils are fair for agricultural use. Most of this soil is timbered with hardwoods.

GRASSLANDS OF WEST CENTRAL UNITED STATES: PRAIRIE STATES TO ROCKY MOUNTAINS

97. ABILENE

Reddish Chestnut – Mollisol – Ustoll – Argiustoll – Pachic Argiustoll.

(Coleman Co., Tex.)

The Abilene series is found largely to the east of the Amarillo soils area of the Southern High Plains. It is much like the

Amarillo, belonging also to a Reddish Chestnut group. In general it is shallower than the Amarillo and occupies much rolling and hilly terrain. It is somewhat darker than the Amarillo and has texture ranging from sandy loam to clay. It is a very productive soil. Much of the rougher terrain is pastured.

98. AMARILLO

Reddish Chestnut – Alfisol – Ustalf – Haplustalf – Typic Haplustalf.

(Randall Co., Tex.)

The Amarillo series is widespread in the Southern High Plains, particularly in the Texas Panhandle and eastern New Mexico. It is a Reddish Chestnut soil formerly classed as Southern Chernozem. It has a reddish or reddish-brown sandy A horizon and a red clay or sandy-clay B horizon. The lime-carbonate layer is 3 to 5 feet beneath the surface. The Amarillo series is one of the most important of the Southern High Plains. Cotton, sorghums, wheat, and some corn are grown. Much of it is in pasture.

99. BARNES

Chernozem– Mollisol – Boroll – Haploboroll – Typic Haploboroll.

(Barnes Co., N.D.)

The Barnes series is the classic example of Chernozem. It occupies wide stretches of flat to gently rolling upland in the Dakotas, extending far into Canada. It is derived from highly calcareous Wisconsin drift with thin loëssal coverings of varying thicknesses. Barnes loam, the dominant type, has a black surface layer 12–15 inches thick, and a brownish subsoil. Distinction between the A and B horizons is hazy. A white, lime layer occupies the lower B horizon and ranges from 3 to 5 feet below the surface. The Barnes is one of the best grain and pasture soils of the region.

100. BATES

Brunizem – Mollisol – Udoll – Argiudoll – Typic Argiudoll.

(Labette Co., Kan.)

The Bates series is associated with Summit soils. They are mainly in eastern Kansas but are also found in northern Oklahoma and western Missouri. The Bates series is derived from noncalcareous shales and sandstones. They have a dark brown granular surface layer and a somewhat heavier-textured brown subsoil. Underneath the subsoil the parent rock is only partially decomposed. Corn, oats, sorghums, and hay crops dominate. Much of the soil is in pasture.

101. BEARDEN

Solonchak – Mollisol – Aquoll – Calciaquoll – Aeric Calci-aquoll.

(Cass Co., N.D.)

The Bearden series, like the Fargo series, is developed from water-laid materials on the bed of old glacial lakes. Ancient Lake Agassiz, now the Valley of the Red River of the North, is the chief area of occurrence. Bearden soil occupies the slightly elevated ridge terrain, largely around the old lake margins. Fargo occupies the flat land. The ridges, in some instances old beach ridges, tend to be sandy or gravelly. Thus Bearden soils are lighter textured than most soils of the area.

102. CARRINGTON

Brunizem – (Not reactivated into the 7th Approximation system).

(Franklin Co., Iowa)

The Carrington series is widely distributed in parts of Minnesota, Iowa, Missouri, Illinois, and Nebraska. It is a maturely developed prairie soil derived from slightly calcareous silty till. It has a relatively thick dark brown surface grading gradually into a dark, somewhat heavier, B horizon. It is thought to be grading toward a Gray-Brown Podzolic group although eluviation is slight. This is a very important agricultural soil. Corn, small grains, and hay crops do well on it.

103. CHEROKEE

Planosol – Alfisol – Aqualf – Albaqualf – Typic Albaqualf.

(Labette Co., Kan.)

The Cherokee series is found in western Kansas, eastern Missouri, and Oklahoma. It is a maturely developed, light-colored, hardpan soil derived from sandstones and shales. It is a relatively infertile soil and is not important agriculturally.

104. CLARION

Brunizem – Mollisol – Udoll – Hapludoll – Typic Hapludoll.

(Faribault Co., Minn.)

The Clarion series is much like the Carrington except that Clarion is poorly drained and not as maturely developed. These soils occur together geographically. Clarion soil is very dark grayish-brown at the surface; the subsoil is distinctly brown. A large proportion of Clarion soil is cultivated. This is an excellent soil for corn and other grains.

105. CLINTON

Gray-Brown Podzolic – Mollisol – Udoll – Hapludoll – Typic Hapludoll.

(Winona Co., Minn.)

The Clinton series is a light-colored, light-textured, maturely developed soil found on the silty loess valley slopes of the Mississippi River and some of its tributaries. It is most extensive in Minnesota, Wisconsin, Illinois, Iowa, and Missouri. The land surface is gentle to sharply rolling and drainage is good. The surface soil is a grayish-brown. The subsoil is of similar color and somewhat heavier. The Clinton series is not an outstanding soil and much of it is in pasture. Crops include corn, small grains, and hay.

106. COLBY

Regosol – Entisol – Orthent – Ustorthent – Typic Ustorthent.

(Custer Co., Neb.)

The Colby series is associated with the Moody series. They are found in the western part of the Corn Belt and are derived from calcareous Peorian loess. They are grayish-brown to brown and sandy or silty in texture. The Colby series has been described as the eroded phase of the Moody series. It is inferior to the Moody for corn and small grain crops.

107. CRETE

Chernozem – Mollisol – Ustoll – Argiustoll – Udic Argiustoll.

(Jefferson Co., Neb.)

The Crete series is found from southern Nebraska to central Kansas. It is associated with Hastings, both being claypan soils. The chief difference is that the Crete series has a higher clay content in its heavy, clay layer. The surface layer is a dark grayish-brown silty clay loam. The upper subsoil consists of a brown, relatively heavy, claypan. A silty, limy material lies below the subsoil. Crete soils are best suited to wheat, other grains, and hay.

108. DANIELS

Chestnut Soil – Mollisol – Ustoll – Argiustoll – Typic Argiustoll.

(Valley Co., Mont.)

The Daniels series is found west of the Williams soils in northeastern Montana. Daniels soils are similar to the Williams, being dark brown with a heavy carbonate zone at a depth of about 2 feet. They differ from the Williams in that they have developed from a thick bed of gravel. Daniels soils are used largely for grazing.

109. DUVAL

Reddish Chestnut – Alfisol – Ustalf – Haplustalf – Udic Haplustalf.

(Hidalgo Co., Tex.)

The Duval series is found on the Rio Grande Plain of southern Texas. Relief is smooth to undulating and the parent material consists of sandy and clayey materials of the Coastal Plain formations. The surface soil is mostly a fine sandy loam, red or reddish-brown. The subsoil is a friable sandy clay. A lime zone in the subsoil is rather weakly developed or missing altogether. Much of the soil is pastured. Dry farming and irrigation both are practiced. Under irrigation, cotton, corn, other feed crops, fruits, and vegetables bear good yields.

110. FARGO

Grumusol – Mollisol – Aquoll – Haplaquoll – Vertic Haplaquoll.

(Traill Co., N.D.)

Fargo soils are associated with the Bearden series, in geographical location and in parent material. Both occupy old lake beds, principally Lake Agassiz, and are formed from waterlaid materials. The Fargo series occupies the flat land of the lake bed. Fargo soils are a black, heavy textured immaturely developed soil, and have only a thin lime layer, which is 2 to 3 feet beneath the surface. These are excellent soils for wheat, other grains, and pasture.

111. GRUNDY

Planosol – Mollisol – Udoll – Argiudoll – Aquic Argiudoll.

(Wayne Co., Iowa)

The Grundy series is a dark prairie soil of eroded loess-covered uplands of northern Missouri and southern Iowa. It is found largely on flat stream divides. It has a hard claypan in the B horizon but it has not developed a lime layer. Intermittently poor drainage has caused the plastic clay subsoil to be mottled in places. The surface layer is dark gray to nearly black, with relatively light texture. The better Grundy soils are excellent for corn and other crops of the area.

112. HASTINGS

Chernozem – Mollisol – Ustoll – Argiustoll – Udic Argiustoll.

(Gage Co., Neb.)

The Hastings soils occupy an area in southern Nebraska and northern Kansas. They are derived from windlaid, calcareous silt materials. These soils are similar to the Moody series, both being lighter in color than the Barnes. The surface layer is light textured. The B horizon is somewhat heavier, having a slight concentration of clay but not a claypan. The zone of lime accumulation is well developed. Hastings soils are used for small grains, alfalfa and other hay crops, and for grazing.

113. HAYS

Reddish Chestnut – (Not reactivated into the 7th Approximation system).

(Saline Co., Kan.)

The Hays series is dominant over a large flat to gently rolling area of central Kansas. These soils are derived largely from shales and limestones, and in the north from Dakota sandstones. They range in texture from sand to clay but they have not developed a claypan. The surface soil is dark grayish-brown and the subsoil dark brown. They become reddish to the south, resembling the Abilene series. The lime layer usually begins at about 30 inches. Dominant uses are for wheat, sorghums, hay, and pasture.

114. HIDALGO

Alluvial – Mollisol – Ustoll – Haplustoll – Typic Haplustoll.

(Hidalgo Co., Tex.)

The Hidalgo series is found in the lower Rio Grande Valley of Texas. They are developed from old alluvial deposits of the broad Rio Grande floodplain. They are relatively young soils and the profile is submature. The subsoil shows a faint accumulation of calcium carbonates. The surface soil is light textured and rich brown or dark brown. They are very productive under irrigation. Cotton, citrus fruits, and vegetables are the main crops.

115. HOLDREDGE

Chernozem – Mollisol – Boroll – Haploboroll – Typic Haploboroll.

(Gosper Co., Neb.)

The Holdredge series is found in southern Nebraska. It is a dark brown sandy to silty loam, and is similar to the Barnes, except that the Holdredge is brown instead of black. Its lime layer lies 2½ to 3 feet beneath the surface. The soil is derived from calcareous loess and is well drained, in spite of level terrain. The Holdredge series is one of the better soils of the area. Corn is the dominant crop, although small grains, sorghums, alfalfa, and other hay crops are important.

116. JOPLIN

Brown Soil – Mollisol – Ustoll – Haplustoll – Typic Haplustoll.

(Washington Co., Colo.)

Joplin soils are found largely on the western side of the Great Plains in Montana, Wyoming, and northern Colorado. Parent material consists of glacial drift in the northern part, while shales and sandstones underlie the central and southern parts. A thin surface layer is gray or grayish-brown, relatively loose. The subsoil is grayish-brown or brown, with a limy layer only 12 to 16 inches below the surface. Most of this soil is pastured.

117. KIRKLAND

Reddish Prairie – Mollisol – Ustoll – Paleustoll – Abruptic Paleustoll.

(Garfield Co., Okla.)

The Kirkland series is found principally in Oklahoma and Kansas. It is a fully mature soil derived from red sedimentary beds. The soil itself is dark brown. The surface layer is light textured but the subsoil is a tough clay with pan-like features dominant. Kirkland soils are used mainly for pasture.

118. MARSHALL

Brunizem – Mollisol – Udoll – Hapludoll – Typic Hapludoll.

(Cherokee Co., Iowa)

The Marshall series is much like the Carrington and somewhat like the Clarion. It is derived from silty, loessal material. It is dark-gray to brown, silty in texture, friable, and highly productive. This is one of the best corn soils in the United States.

119. MILES

Reddish Chestnut – Alfisol – Ustalf – Haplustalf – Udic Haplustalf.

(Childress Co., Tex.)

The Miles series is found associated with similar soils over a large area in the rolling plains section of northern and northwestern Texas, western Oklahoma, and southern Kansas. They frequently are badly eroded. They are typical of the red lands of the area and are derived from clays and shales of the "Red Bed" formations. Surface soils range from red to reddish-brown, with texture from sand to clay. The lime layer in the subsoil is typical of the area except that it is largely lacking in the badly eroded steeper sections. Often these soils are relatively deep, friable, and productive of cotton, corn, sorghums, and small grains. Much of the Miles soils are used for grazing.

120. MILLER

Alluvial – Mollisol – Ustoll – Haplustoll – Vertic Haplustoll.

(Cleveland Co., Okla.)

The Miller series is a reddish-brown soil of the Texas Panhandle. It is derived from stream alluvium and is not very extensive, since floodplains are narrow and widely spaced. The Miller series is associated geographically with the upland Ama-

rillo series. They are good producers of the crops common
to the area.

121. MOODY

Chernozem – Mollisol – Ustoll – Haplustoll – Udic Haplustoll.
(Cedar Co., Neb.)

The Moody series occupies a relatively small area in the western part of the Corn Belt in northeastern Nebraska and con-
tiguous states. These soils are derived from calcareous loess
of Peorian age. They are grayish-brown to almost black surface soils, grading downward to almost black, friable, silty
subsoils. A well-marked lime layer begins at 18 to 30 inches.
A layer of lime concretions in the upper part of the lime
layer distinguishes the Moody series. The soil is excellent for
corn and wheat.

122. PARSONS

Brunizem – Alfisol – Aqualf – Albaqualf – Mollic Albaqualf.
(Labette Co., Kan.)

The Parsons series is associated with Grundy and Shelby soils
in southern Iowa, northern Missouri, and eastern Kansas. All
three occupy a loess-covered plain. The Parsons soils are
largely in the southern part of the area. They have dark,
grayish-brown surface layers and heavy, claypan subsoils. Parsons soils are poorly drained and are only moderately productive. Many have been drained, limed, and fertilized. Corn is
the leading crop.

123. PHILLIPS

Sierozem – Aridisol – Argid – Natrargid – Glossic Mollic Natrargid.
(Phillips Co., Mont.)

A belt of Phillips soils lies along the Canadian border in
Montana just north of the Missouri River. This area contains

a very large number of "slick spots," or Solonetz areas. These elliptical areas are small depressions less than 50 feet in diameter and 6 inches deep. Sometimes they are called "buffalo wallows." They give an area a pockmarked appearance, hence the term, "Smallpox of the Plains." These soils have a heavy, tough, clay horizon near the surface. The parent material is very low in sodium carbonate. Phillips soils are of very low value.

124. PIERRE

Brown Soil – Aridisol – Orthid – Camborthid – Ustertic Camborthid.

(Butte Co., S.D.)

The Pierre series is characteristic of western South Dakota, extending into Nebraska, Wyoming, and Montana. It is derived from Pierre shales and other similar formations. Surface soils are moderately dark and dense, and subsoils are clayey or shale. The soil is extremely sticky and plastic when wet, and hard and tough when dry. The lime layer occurs at or near the surface. Wheat and other small grains are the principal crops, but light rainfall keeps yield low. Much of the soil is pasture.

125. PUTNAM

Planosol – Alfisol – Aqualf – Albaqualf – Mollic Albaqualf.

(Christian Co., Ill.)

The Putnam series is developed from silty loessal material. The main area of its occurrence is parts of southern Ohio, Indiana, Illinois, and northeastern Missouri. Much of the area has Illinoian till underneath the loess. The soils develop under conditions of permanently poor or seasonally poor drainage, and subsoils are heavy, tough, and impervious. Claypans are common. Surface soils are silty, gray to dark gray loams. Subsoils are heavy and mottled. Yields are low. Corn, small grains, and hay are the principal crops.

126. ROSEBUD

Chestnut Soil – Mollisol – Ustoll – Argiustoll – Typic Argiustoll.

(Wibaux Co., Mont.)

The dark brown Rosebud series is a dominant soil of an area in western Nebraska and adjacent parts of adjoining states. The soil occupies flat to undulating divides. The area is unglaciated and soils are derived from light-colored calcareous shales and sandstones. The dominant type is a fine sandy loam. It is brown colored and light textured, with subsoil only slightly heavier than the surface layer. Smoother areas of Rosebud soil are used for wheat, corn, and oats. Much of it is pasture.

127. SHELBY

Brunizem – Mollisol – Udoll – Argiudoll – Typic Argiudoll.

(De Kalb Co., Mo.)

The Shelby soil is associated with the Grundy series in southern Iowa and northern Missouri. They have developed on a loess-covered plain and have a grayish-brown surface layer and a brown or yellowish-brown gritty subsoil. Below the subsoil the material is glacial till. Shelby soils are inferior to Grundy soils. Since they occur on steep terrain, often they are not suited to cultivation. Crops consist of corn, oats, and hay. Much of the soil is pastured.

128. SUMMIT

Brunizem – Mollisol – Udoll – Argiudoll – Vertic Argiudoll.

(Lafayette Co., Mo.)

The Summit series is the typical mature soil derived from unglaciated, slightly calcareous shales, sandstones, and limestones. It is a significant grassland soil found in western Missouri and eastern Kansas. It is dark brown to black and is a good grazing and grain growing soil.

129. TAMA

Brunizem – Mollisol – Udoll – Argiudoll – Typic Argiudoll.

(Crawford Co., Wis.)

The Tama series is a dark-colored prairie soil with maturely developed profile. It is derived from old, leached loëss. It is widely distributed in Iowa, Illinois, Nebraska, and southern Wisconsin. The soil has good tilth, is relatively productive, and presents a pleasing undulating to rolling landscape. It is one of the Corn Belt's most attractive soils. The Tama series is associated with the Marshall. The Tama is a deep, mellow soil. Corn and oats are dominant crops.

130. VALERA

Reddish Chestnut – Mollisol – Ustoll – Calciustoll – Petrocalcic Vertic Calciustoll.

(Coleman Co., Tex.)

The Valera soil is best developed in the Edwards Plateau of Texas. It is a dark, shallow soil about one foot in total thickness with a distinct caliche layer underneath the thin, often stony, surface layer. Most land of this area is used for pasture for sheep and goats.

131. VICTORIA

Grumusol – Vertisol – Ustert – Pellustert – Typic Pellustert.

(Cameron Co., Tex.)

The Victoria series occupies a limited portion of the lower Rio Grande Plain region. The surface soil is granular, dark brown or black, with a yellowish-brown to dark-brown clay subsoil. The soil is derived from calcareous clays of Coastal Plain deposits. They formerly were classed as Rendzinas and are highly calcareous. These soils are inherently fertile but crops often suffer from insufficient moisture. Irrigation brings good crop yields. However, much of the soil is pastured.

132. WEBSTER

Humic-Gley – Mollisol – Aquoll – Haplaquoll – Typic Haplaquoll.

(Webster Co., Iowa)

Webster soils are found principally in the four states: Indiana, Illinois, Iowa, and Minnesota. They are prized corn soils and are derived from calcareous glacial till. They are associated with Clarion soils, but are heavier textured, blacker, and not as well drained as the Clarion, since the Webster occupies the flatter areas. The surface soils are light textured and the subsoil is mottled. Corn is the dominant crop, with oats and soy beans following. Yields are very high.

133. WILLIAMS

Chestnut Soil – Mollisol – Ustoll – Argiustoll – Typic Argiustoll.

(Potter Co., S.D.)

The Williams series borders the Barnes series on the west. It is derived from thinner deposits of calcareous till, and because of less luxuriant grass cover, is lighter in color than most Chernozems further east. The rather thin surface soil—6 to 10 inches—is dark gray (not black like the Barnes). The brownish B horizon also is thin, with the lime layer about 18 inches beneath the surface. The texture usually is sandy or gravelly and the material is friable. These are dominantly wheat and grazing soils.

WESTERN UNITED STATES: ROCKY MOUNTAINS TO THE PACIFIC

134. EVERETT

Gray-Brown Podzolic – Spodosol – Orthod – Haplorthod – Typic Haplorthod.

(King Co., Wash.)

The Everett series is an important soil in the Puget Sound

Lowland of Washington. It belongs to the Gray-Brown Podzolic group, has developed under heavy forest, and is only mildly acid. The parent material is glacial till of heterogeneous rock origin. Outwash and terrace materials also are included. Everett soils are light-colored, sandy, gravelly, and stony loams. These soils are used largely for timber and recreation. High clearing costs discourage agricultural uses. Dairying and general farming are of some importance.

135. FRESNO

Sierozem – Aridisol – Argid – Durargid – Typic Durargid.

(Fresno Co., Calif.)

Fresno soils occur in dry interior valleys of California, mainly in the San Joaquin Valley. They are light brown or light grayish-brown soils varying in texture from sand to clay. The subsoil is a gray limy claypan. Some of the Fresno soils are saline Solonchak; however, the sandier types, which are better drained, contain little salt. The better drained, sandy, salt-free soils are intensively cultivated, largely under irrigation. Cotton, alfalfa, grapes, peaches, and vegetable crops are widely grown and yield heavily.

136. HANFORD

Alluvial – Entisol – Orthent – Xerorthent – Typic Xerorthent.

(Los Angeles Co., Calif.)

The Hanford series is a young, dark brown soil found on stream deltas and alluvial fans in southern California. The chief areas are in the mountain foothills east of the San Joaquin Valley and in the Los Angeles area. These soils do not have a developed profile and there is no zone of lime accumulation. Hanford soils are among the most productive of the region. They are well drained, free of salt, and have a friable, sandy to loamy texture. Most of those under cultivation are irrigated. A variety of field crops, citrus and deciduous fruits, grapes, and vegetables yield well on Hanford soils.

137. HELMER

Regosol – Entisol – Orthent – Xerorthent – Aquic Xerorthent.

(Latah Co., Idaho)

The Helmer series is distributed widely in a large area in northern Idaho, western Montana, Wyoming, Colorado, and in the Sierra Nevadas of California. It has developed from crystalline bedrock under forest vegetation. They are shallow, grayish-brown soils, being granular at the surface and more compact below. Profiles are weakly developed but show evidence of Podzolization. Helmer soils have slight agricultural value and only limited grazing value. Timber growing and recreation are the chief uses.

138. IMPERIAL

Alluvial – Entisol – Orthent – Terriorthent – Typic Terriorthent.

(Imperial Co., Calif.)

The Imperial series is the dominant soil of a portion of the desert of southern California. It is not a Desert soil in a technical sense, since it is derived from very recently deposited alluvial materials brought in by the Colorado River in flood. It is surrounded largely by Mohave and related series of true desert soils. The Imperial is a brown to dark brown, highly fertile, alluvial soil. It and its associates form the basis of the very affluent agriculture of the Imperial Valley. Fruits, vegetables, cotton, sugar beets, and other crops are grown.

139. MADERA

Noncalcic Brown – Alfisol – Xeralf – Durixeralf – Typic Durixeralf.

(Madera Co., Calif.)

The chief area of Madera soil is in the central and northern parts of the Central Valley of California. The soil occupies gently sloping valley plains and stream terraces. Parent material is alluvial deposits. Drainage is usually ample, but slope is slight, and much runoff is into locally depressed areas where drainage is slow. Surface soils are rich brown to reddish-brown. Subsoils are heavier textured, compact, and plastic, with a brown to red iron-and-silica-cemented hardpan. The Madera soil is used for general farming, dairying, and growing fruit orchards.

140. MELBOURNE

Reddish-Brown Lateritic – Ultisol – Umult – Haplohumult – Xeric Haplohumult.

(Clatsop Co., Ore.)

The Melbourne series is found in the coastal mountain and hill lands of western Washington, western Oregon, and north-western California, and in the western foothills of the Cascades. They are dark brown, friable soils becoming moderately compact and plastic in the subsoil. They are derived from shales and sandstones and have developed under forest vegetation. Terrain varies from low valley slopes to rugged hill land. These soils are used largely for forests and pastures, with some orchards at lower elevations.

141. MOHAVE

Red Desert – Aridisol – Argid – Haplargid – Typic Haplargid.

(Maricopa Co., Ariz.)

The Mohave desert soil is a dominant soil over a large portion of the cactus-studded desert of southern Nevada, southern California, southwestern Arizona, New Mexico, and Texas. It is a reddish desert soil developed from water-worked materials of alluvial fans, stream terraces, and plains. The surface is sandy or silty, sometimes containing salt patches. The subsoil is reddish-brown or red, heavy, tough clay. Much of this soil is used for grazing. Cultivation requires irrigation and drainage. Citrus fruits, winter vegetables, and cotton yield well on the better soils where water supply is adequate.

142. NEZ PERCE

Planosol – Mollisol – Xeroll – Palexeroll – Abruptic Palexeroll.

(Nez Perce Co., Idaho)

The Nez Perce soil area is located near the mountains in Idaho, eastern Oregon, and eastern Washington in the loess-covered Columbia Plateau wheat region. The soil is a grassland Pedocal, although the lime layer usually is not present within 6 feet of the surface. Surface soils are dark grayish-brown or black, granular silt loams or clay loams. The subsoil is yellowish-brown clay loam or clay. Most Nez Perce soils are dry-farmed. Wheat is the chief crop, but hay, peas, beans, and potatoes are also planted.

143. OLYMPIC

Reddish-Brown Lateritic – Ultisol – Umult – Haplohumult – Xeric Haplohumult.

(Clallam Co., Wash.)

The Olympic series dominates the western slopes of the Cascade Mountains in Washington and Oregon. It also is important in the coastal area from the Olympic Mountains southward. This soil is derived principally from basaltic rocks and volcanic tuffs, and is acid in reaction. Surface soils are dark brown clay loams. Subsoils are brown to reddish. The soils are largely devoted to forestry and grazing, but a limited proportion of them are devoted to fruits, walnuts, and berries.

144. PALOUSE

Brunizem – Mollisol – Xeroll – Haploxeroll – Typic Haplo-xeroll.

(Spokane Co., Wash.)

Palouse soils occupy rolling uplands in the loess-covered Columbia Plateau wheat section of Washington, Oregon, and Idaho. The area usually is designated as Palouse country. Surface soils are dark grayish-brown to black, granular silt loams or loams. Subsoils are lighter colored but somewhat heavier textured. A lime carbonate layer may or may not be concentrated within 6 feet of the surface. Palouse soils are very fertile, producing high wheat yields by fallowing every second or third year. Beans, alfalfa, other hay crops, and potatoes yield well.

145. POND

Sierozem – Aridisol – Argid – Haplargid – Mollic Haplargid.

(Los Angeles Co., Calif.)

The Pond series is associated with the Fresno soils of the San Joaquin Valley of California. Both are claypan soils, but they differ in that the Pond series is distinctly saline, with salt crusts and small "slick spots" of Solonetz (alkali claypan). Pond soils are not productive unless drained and washed of salt and alkali.

146. PORTNEUF

Sierozem – Aridisol – Orthid – Calciorthid – Mollic Calciorthid.

(Bingham Co., Idaho)

The Portneuf series is found in a relatively large intermountain desert plain within the Great Basin province, largely in southern Idaho and adjacent parts of Utah and Oregon. Surface soils of this series range from light gray to brown, with textures ranging from sand to clay loam. Subsoils are light gray, compact, and distinctly lime. Most of this soil is devoted to grazing cattle and sheep. Its carrying capacity is relatively low.

147. RAMONA

Noncalcic Brown – Alfisol – Xeralf – Haploxeralf – Typic Haploxeralf.

(San Diego Co., Calif.)

The Ramona series is restricted to alluvial fan and old stream terrace areas largely in southern California and to a limited degree in central California. The alluvial fan and valley materials are derived from granitic rocks. Soils are reddish-brown, gritty sandy loam at the surface, with compact, moderately heavy-textured subsoils. Uses of these soils present wide contrasts. Citrus grows in some, deciduous fruits in others, vegetables in still others, all by irrigation. Some nonirrigated Ramona soils are planted in grains and hay crops.

148. REEVES

Red Desert – Aridisol – Orthid – Calciorthid – Typic Calciorthid.

(Reeves Co., Tex.)

The Reeves series is a desert soil found in a large region extending from western Texas to California. It is associated with the Mohave series. Both are derived from water-worked materials on alluvial fans, terraces, and stream plains. The Reeves differs from the Mohave in that the Reeves tends toward a more highly calcareous subsoil with a cemented hardpan of lime, iron, and silica and with occasional concentrations of gypsum. This soil is used largely for grazing. Cropgrowing by irrigation is limited.

149. RITZVILLE

Brown – Mollisol – Xeroll – Haploxeroll – Calcic Haploxeroll.

(Franklin Co., Wash.)

The Ritzville soil occupies a small region on the Columbia Plateau in Idaho, eastern Washington, and eastern Oregon, west of the belt of Palouse soils. The soil is derived from fine loess. The surface is a mellow, light grayish-brown, fine sandy loam to silt loam. The pervious subsoil is lighter colored than the surface soil. A distinct layer of lime accumulation is at depths of 3 to 6 feet. The greater part of Ritzville soil is dry-farmed to wheat. Yields are moderate. Much of this soil is devoted to cattle and some sheep grazing.

150. SAN JOAQUIN

Noncalcic Brown – Alfisol – Xeralf – Durixeralf – Abruptic Durixeralf.

(Merced Co., Calif.)

The San Joaquin series is a dominant soil of central California. The main area of occurrence is east of the San Joaquin and lower Sacramento Rivers midway between the rivers and the Sierra Nevadas. Surface soils are rich brown to reddish-brown, light textured, and plastic, resting on brown to red cemented iron-and-silica hardpans. The entire solum is acid. Most of these soils are dry-farmed in wheat, barley, and pasture. Some citrus fruits and figs are grown.

151. SIERRA

Noncalcic Brown – Alfisol – Xeralf – Haploxeralf – Ultic Haploxeralf.

(Fresno Co., Calif.)

The Sierra series is one of several soils occupying the lower Sierra Nevada foothill belt in southern Calfornia and in southern Arizona. Slopes are gentle to moderate but rock outcrops are numerous. These soils are derived from granite parent materials. Surface soils are rich brown to reddish-brown, gritty and granular sandy loams. Subsoils are brown or red, compact and blocky in structure. Sierra soils are mildly acid. Bedrock is at depths ranging from a few inches to several feet. When irrigated, some of these foothill soils are highly productive of deciduous and citrus fruits.

152. SPRINGDALE

Brown Forest – Mollisol – Xeroll – Haploxeroll – Entic Haploxeroll.

(Bonner Co., Idaho)

The Springdale series is found in the Puget Sound Lowland of Washington and in eastern Washington and western Idaho. They are everywhere associated with Everett soils but are less extensive in the Puget Sound area and reach their greatest area in eastern Washington and Idaho. Springdale soils have somewhat sandier subsoils than the Everett soils; otherwise they are quite similar. Like the Everett, Springdale soils are not highly important, agriculturally.

153. WILLAMETTE

Gray-Brown Podzolic – Mollisol – Xeroll – Argixeroll – Pachic Ultic Argixeroll.

(Yamhill Co., Ore.)

The Willamette series is the dominant soil of the Willamette Valley of western Oregon and an outlying area of southwestern Washington. It is a maturely developed soil derived from water-laid materials laid down in the process of valley filling. Willamette soils are mildly to moderately acid and have no lime layer. Surface soils are light gray to rich brown loams or clay loams. Brown subsoils are permeable and mellow. The Willamette is the most important soil of the area. General farming, dairying, truck farming, and fruit farming are well developed.

UNITED STATES SOILS OUTSIDE THE FORTY-EIGHT CONTERMINOUS STATES

ALASKA

154. FAIRBANKS

Subarctic Brown Forest – Inceptisol – Ochrept – Cryochrept – Typic Cryochrept.

(Fourth Judicial Division, Alaska)

The Fairbanks is one of the dominant soils of Alaska. These are forested soils that occupy well-drained slopes above the river bottoms (occupied by Tanana soils) in the interior of Alaska above the junction of the Yukon and Tanana rivers. The dominant type is a deep, mellow brown silt loam underlain by yellowish-brown, moderately compact silt loam subsoil. This soil is derived from loess or from a mixture of loess and water-laid material. Fairbanks soils are not farmed extensively, a small fraction of the total one-half million acres being in crops. Potatoes, small grains, and vegetables yield quite well.

155. YUKON

Low-Humic Gley – Inceptisol – Aquept – Cryaquept – Histic Pergelic Cryaquept.

(Fourth Judicial Division, Alaska)

The Yukon series is associated with the Fairbanks in the interior of Alaska. They are largely restricted to the Yukon Valley and are derived from alluvial materials. They are brown forest soils and are less acid than other alluvial soils of the area. They are not widely utilized, but they are considered to be suited to general farming, dairying, and cattle grazing.

HAWAII

156. HONOULIULI

Low-Humic Gley – Inceptisol – Tropept – Ustropept – Vertic Ustropept.

(Oahu, Hawaii)

The Honouliuli series is found on the Ewa coastal plain surrounding Pearl Harbor on the Hawaiian island of Oahu. The soil is derived from marine clays that have recently been deposited. It is a very young, heavy-textured soil with scarcely any profile features. It is used to some extent for truck crops and fruits.

157. MAMALA

Lithosol – Inceptisol – Tropept – Ustropept – Lithic Ofic Ustropept.

(Oahu, Hawaii)

The Mamala series is found in the Pearl Harbor area—the Ewa coastal plain of the Hawaiian island of Oahu. It is associated with the Honouliuli series in that it lies on the Coral Plain surrounding the clay beds on which the Honouliuli soil is located. The Mamala soil is one of the few Hawaiian upland soils that is not derived from volcanic materials. Derived from coral, it is neutral or only slightly acid. Its profile is not maturely developed and the distinction between horizons is somewhat hazy. This soil is not widespread. It is fair for sugar cane, vegetables, and fruits.

PANAMA CANAL ZONE

158. FRIJOLES

Latosol – (Not reactivated into the 7th Approximation system).

(Canal Zone)

The Frijoles series is one of the dominant soils of the Panama Canal Zone. It is a typical soil of tropical savanna lands that have distinct wet and dry seasons. It is a deep red clay soil, slightly brownish at the surface and brick red in the subsoil. It is derived from dark-colored basic igneous rocks and is more friable than most middle latitude clays. Principal crops are bananas, pasture grass, rice, vegetables, papayas, and mangoes.

PUERTO RICO

159. CATALINA

Latosol – Oxisol – Orthox – Haplorthox – Tropeptic Haplorthox.

(Ponce Senatorial District, Puerto Rico)

The Catalina series is a prominent Latosol soil of Puerto Rico. This is a brownish-red or red, slightly granular clay or silty clay. It is well drained, permeable, and moderately acid. It is derived from tuffaceous rocks, shales (from volcanic ash), conglomerates, and other rocks. Weathering is rapid and relatively deep. They are not highly susceptible to erosion and relatively steep slopes are cultivated safely. Catalina soils are suited to a variety of tropical crops, but they are inclined to exhaust plant nutrients and require considerable fertilizing. Coffee, oranges, and bananas are principal crops.

160. DESCALABRADO

Lithosol – Mollisol – Ustoll – Haplustoll – Lithic Haplustoll.

(Guayama Senatorial District, Puerto Rico)

The Descalabrado series is found in arid to subhumid areas of Puerto Rico. It is a Lithosol derived from tuffaceous rocks and shales. The soil has a brown or dark brown, granular-clay or silty clay-loam surface layer, which is neutral or alkaline in reaction. The subsoil is light brown silty clay, also neutral or calcareous. The soil is shallow; the bedrock usually ranges in depth from 8 to 18 inches. This is primarily a pasture soil. Grass grows well and cures well. Its capacity for grazing cattle is relatively high.

161. NIPE

Latosol – Oxisol – Orthox – Acrorthox – Typic Acrorthox.

(Mayaguez Senatorial District, Puerto Rico)

Nipe soils are deep, purplish-red permeable clays, high in iron and aluminum. They are Latosols, high in chromium and nickel oxides and low in silicon dioxide. The surface soil is high in organic matter, the subsoil quite low. Weathering is deep, and bedrock does not occur above a depth of as much as 20 feet in most places. The parent material is serpentine. Water penetration is rapid. Agricultural use is restricted to a few areas where heavy fertilization can be practiced.

162. PONCENA

Grumusol – Vertisol – Ustert – Pellustert – Chromudic Pellustert.

(Guayama Senatorial District, Puerto Rico)

The Poncena series is one of several Grumusols in southern Puerto Rico. It occupies level to undulating terrain and is derived from shaly limestones and tuffaceous rocks and in part from similar materials washed down from the hills. These soils are well suited to sugar cane and good pasture grasses. Some irrigation is required for sugar cane. The soils lie well, are easily cultivated, have good depth, and have the capacity to retain water.

163. SOLLER

Rendzina – Mollisol – Endoll – Rendoll – Entropeptic Rendoll. (Mayaguez Senatorial District, Puerto Rico)

The Soller soils are Rendzinas occupying belts of undulating or hilly coastal plains along both northern and southern coasts of Puerto Rico and neighboring islands. They are dark soils underlain by limestones. Surface streams are lacking. The Soller soils are in the rainier portions of the coastal plains. Surface soils range in depth between 3 and 20 inches. They consist of yellowish-brown, silty, somewhat plastic, calcareous material, underlain by white limestone or marl. Grasses and other crops are grown but yields are low.

Useful References

BOOKS

Bear, Firman E. *Earth—The Stuff of Life*. Norman, Oklahoma: University of Oklahoma Press, 1961.

Bennett, Hugh Hammond. *Elements of Soil Conservation*. New York: McGraw-Hill Book Company, 1955.

———— and William Clayton Pryor. *This Land We Defend*. New York: Longmans, Green and Company, 1942.

Black, C. A. *Soil Plant Relationship*. New York: John Wiley, 1957.

Brewer, Roy. *Fabric and Mineral Analysis Soils*. New York: John Wiley, 1964.

Buckman, Harry J. *The Nature and Properties of Soils*. New York: Macmillan, 1960.

Bunting, Brian Talbot. *The Geography of the Soil*. Chicago: Aldine Publishing Company, 1965.

Bushnell, Thomas Mark. *A Story of Hoosier Soils*. Lafayette, Indiana: Peda Products, 1958.

————. *The Story of Indiana Soils*. Lafayette, Indiana: Purdue University Experiment Station, 1944.

Clarke, George Robin. *The Study of the Soil in the Field*. Oxford, England: The Clarendon Press, 1961.

Colby, C. B. *Soil Savers*. New York: Coward-McCann, Inc. 1957.

Donahue, Roy Luther. *Our Soils and Their Management*. Danville, Illinois: Interstate Printers and Publishers, 1961.

Eyre, S. R. *Vegetation and Soils*. Chicago: Aldine Publishing Company, 1964.

Faulkner, Edward H. *Soil Development*. Norman, Oklahoma: University of Oklahoma Press, 1952.

Great Plains Committee. *The Future of the Great Plains*. Washington, D.C.: United States Government Printing Office, 1936.

Held, R. Burnell, and Marion Clawson. *Soil Conservation in Perspective*. Baltimore, Maryland: The Johns Hopkins Press, 1965.

Hilgard, Eugene Waldemar. *Soils, Their Formation, Properties, Composition, and Relation to Climate and Plant Growth in the Humid and Arid Regions*. New York: The Macmillan Company, 1921.

Jacks, G. V. *Vanishing Lands*. New York: Doubleday, Doran and Company, 1939.

Jackson, Marion LeRoy. *Soil Chemical Analysis*. Englewood Cliffs, New Jersey: Prentice-Hall, 1958.

Joffe, J. S. *Pedology*. New Brunswick, New Jersey: Rutgers University Press, 1936.

Joly, John. *Surface History of the Earth*. Oxford, England: The Clarendon Press, 1930.

Kellogg, Charles Edwin. *The Soils That Support Us*: *An Introduction to the Study of Soils and Their Use by Man*. New York: The Macmillan Company, 1941.

Lee, William D. *The Soils of North Carolina*: *Their Formation, Identification, and Use*. Raleigh, North Carolina: North Carolina Agricultural Experiment Station, Technical Bulletin Number 115, 1955.

Lord, Russell. *Behold Our Land*. Boston: Houghton Mifflin, 1938.

Marbut, Curtis F. Memorial Volume by Committee. *Soils: Their Genesis and Classification*. Soil Science Society of America, 1951.

Marshall, Charles Edmund. *The Physical Chemistry and Mineralogy of Soils*. New York: John Wiley, 1964.

Mickey, Karl B. *Man and the Soil.* Chicago: International Harvester Company, 1945.

Miller, Charles E. *Fundamentals of Soil Science.* New York: John Wiley, 1958.

Parkins, A. E., and J. R. Whitaker. *Our National Resources and Their Conservation.* New York: John Wiley and Sons, Inc., 1936.

Poirot, Eugene M. *Our Margin of Life.* New York: Vantage Press, 1964.

Renne, Roland R. *Land Economics.* New York: Harper and Brothers, 1958.

Robinson, Gilbert Wooding. *Soils, Their Origin, Constitution and Classification: An Introduction to Pedology.* London, England: T. Murby, 1949.

Schneider, Herman, and Nina Schneider. *Rocks, Rivers, and the Challenging Earth.* New York: William R. Scott, Inc., 1952.

Sears, Paul B. *Deserts on the March.* Norman: University of Oklahoma Press, 1935.

Shaw, Byron T. *Soil Physical Conditions and Plant Growth.* New York: Academic Press, 1952.

Shepard, Jean Key. *Food or Famine.* New York: The Macmillan Company, 1945.

Sherman, Robert Clyde. *Life and Death of the Soil.* Chicago: Science Research Associates, 1955.

Soil – The Yearbook of Agriculture. Washington, D. C.: The United States Department of Agriculture, 1957.

Staff, Soil Science Society of America. Memorial Volume. *Life Works of Curtis F. Marbut.* Soil Science Society of America, 1942.

Stallings, J. H. *Soil Conservation.* Englewood Cliffs, N. J.: Prentice-Hall, Inc., 1957.

————. *Soil–Use and Improvement.* Englewood Cliffs, N. J.: Prentice-Hall, Inc., 1957.

Tedrow, J. C. F. *Antarctic Soils*. American Geophysical Union. N.A.S., 1966.

Teuscher, H., and R. Adler. *The Soil and Its Fertility*. New York: Reinhold Publishing Corporation, 1960.

U. S. Department of Agriculture. *Agricultural Land Resources*. Bulletin 263. Washington, D. C.: U. S. Government Printing Office, 1962.

————. Agricultural Research Service. *Water Intake by Soil*. Bulletin No. 925. Washington, D. C.: U. S. Government Printing Office, 1963.

————. Soil Conservation Service. *That Land Down There*. Bulletin No. 255. Washington, D. C.: U. S. Government Printing Office, 1962.

————. *Soil Conservation at Home*. Bulletin No. 244. Washington, D. C.: U. S. Government Printing Office, 1962.

————. *Soil Erosion*. Bulletin 260. Washington, D. C.: U. S. Government Printing Office, 1962.

————. *Soils and Men—The Yearbook of Agriculture*. Washington, D. C.: U. S. Government Printing Office, 1938.

————. *Wind Erosion and Dust Storms*. Leaflet No. 394. Washington, D. C.: U. S. Government Printing Office, 1961.

Welch, Charles D., and Gerald D. McCarty. *An Introduction to Soil Science in the Southeast*. Chapel Hill, North Carolina: The University of North Carolina Press, 1963.

Whitney, Milton. *Soil and Civilization*. New York: D. Van Nostrand Company, 1925.

PERIODICAL ARTICLES

Annals of Association of American Geographers
    Marbut, Curtis F. "Soils of the Great Plains." (Vol. 13, 1923).
Economic Geography
    Gibson, J. Sullivan. "The Soils Factor in the Character of Land Use in the Tennessee Valley." (Vol. 13, 1937).

Geographical Review
   Wolfanger, Louis. "Major World Soil Groups." (Vol. 19, 1929).
   _____. "Economic Geography of Great Soil Groups of Eastern United States." (Vol. 21, 1931).
Occasional Papers in Geography and Geology
   Indiana State University, Terre Haute, Indiana
      Gibson, J. Sullivan. "Selections of Significance to the Geographer, from Soil Science Society of America Proceedings." (Paper number one, 1968).
Soil Science of America—Proceedings.
      Ableiter, J. K. "Productivity Ratings in the Soil Survey Reports." (Vol. 2, 1937, pp. 415–422).
      Andrew, L. E., and H. F. Rhodes. "Soil Development from Calcareous Glacial Material in Eastern Nebraska during Seventy-Five Years." (Vol. 12, 1947, pp. 407–408).
      Austin, M. E. "Report of Committee on Exchange of Soil Pictures and Soil Profiles." (Vol. 7, 1942, pp. 454–459).
      Baxter, F. P., and F. D. Hole. "Ant (*Formica Cinerea*) Pedaturbation in a Prairie Soil." (Vol. 31, 1967, pp. 425–428).
      Bushnell, T. M. "Some Aspects of the Catena Concept." (Vol. 7, 1942, pp. 466–476).
      Chandler, R. F. "The Time Required for Podzol Profile Development as Evidenced by the Mendenhall Glacier Deposits Near Juneau, Alaska." (Vol. 7, 1942, pp. 454–459).
      Coile, T. S. "Podzol Soils in the Southern Appalachian Mountains." (Vol. 3, 1938, pp. 274–279).
      Daniels, R. B., W. D. Nettleton, R. J. McCracken, and E. E. Gamble. "Morphology of Soils with Fragipans in Parts of Wilson County, North Carolina." (Vol. 30, 1966, pp. 376–380).

Ekblaw, W. E. "Soil Science and Geography." (Vol. 1, 1936, pp. 1–5).

Hester, J. B. "The Relation of Soil Texture and Color to the Organic Matter Content." (Vol. 3, 1938, pp. 112–114).

Holmes, R. S. "Influence of Drainage Upon Coastal Plain Soils." (Vol. 1, 1936, pp. 161–163).

Joffe, J. S. "Soil-Forming Processes: Pedology in the Service of Soil Science." (Vol. 6, 1949, pp. 68–77).

Kellogg, C. E. "The Future of the Soil Survey." (Vol. 14, 1949, pp. 8–13).

—————. "Soil Classification and Cartography in Relation to Other Soil Research." (Vol. 4, 1939, pp. 339–342).

Klingebiel, A. A. "Soil Survey Interpretation: Capability Groupings." (Vol. 21, 1958, 160–163).

Krusekopf, H. H. "The Hardpan Soils of the Ozark Region." (Vol. 7, 1942, pp. 434–436).

Lapham, M. H. "The Soil Survey from the Horse-and-Buggy Days to the Modern Age of the Flying Machine." (Vol. 10, 1945, pp. 334–350).

Leyford, W. H., Jr. "The Morphology of the Brown Podzolic Soils of New England." (Vol. 11, 1946, pp. 486–492).

—————. "Characteristics of Some Podzolic Soils of the Northeastern United States." (Vol. 16, 1952, pp. 231–235).

Matthews, H. L., G. W. Prescott, and S. S. Obenshain. "The Genesis of Certain Calcareous Floodplain Soils of Virginia." (Vol. 29, 1965, pp. 729–732).

Matzek, B. L. "Movement of Soluble Salts in Development of Chernozems and Associated Soils." (Vol. 19, 1955, pp. 225–229).

Morgan, C. G., and S. S. Obenshain "Genesis of the Soils Developed from Materials Residual from Limestone." (Vol. 7, 1942, pp. 441–447).

Norton, E. A. "Classes of Land According to Use Capabilities." (Vol. 4, 1939, pp. 378–380).

Oakes, H., and J. Thorp. "Dark Clay Soils of Warm Regions Variously Called Rendzinas, Black Cotton Soils, Regur and Tirs." (Vol. 15, 1950, pp. 347–354).

Peele, C. T. "The Effect of Calcium on the Erodibility of Soils." (Vol. 1, 1936, pp. 47–51).

Rice, T. D. "Physical Characteristics of the Soil Profile as Applied to Land Classification." (Vol. 1, 1936, pp. 455–458).

Ruhe, R. V., J. M. Williams, R. C. Shuman, and E. L. Hill. "Nature of Soil Parent Materials in Ewa-Waipahu Area, Oahu, Hawaii." (Vol. 29, 1965, pp. 282–287).

Simonson, R. W. "Studies of Buried Soils Formed from Till in Iowa." (Vol. 6, 1941, pp. 373–381).

——————. "Genesis and Classification of Red-Yellow Podzolic Soils." (Vol. 14, 1949, pp. 316–319).

Soileau, J. M., and R. J. McCracken. "Free Iron and Coloration in Certain Well-Drained Coastal Plain Soils in Relation to their Other Properties and Classification." (Vol. 31, 1967, pp. 248–254).

Storie, R. E. "Soil Regions of California Illustrated by Twenty-Four Dominant Soil Types." (Vol. 11, 1946, pp. 425–430).

Stobbe, P. C. "The Morphology and the Genesis of the Gray-Brown and Related Soils of Eastern Canada." (Vol. 16, 1952, pp. 81–84).

Sturgis, M. B., and C. W. McMichael. "The Genesis and the Morphology of the Soils of the Lower Mississippi Delta." (Vol. 4, 1939, pp. 358–359).

Templin, E. H., and G. W. Kanze. "Houston Black Clay, the Type Grumusol: II Mineralogical and Chemical Characterization." (Vol. 20, 1956, pp. 91–96).

Thorp, J. "The Influence of Environment on Soil Formation." (Vol. 6, 1941, pp. 39–46).

—————, B. H. Williams, and W. I. Watkins. "Soil Zones of the Great Plains States—Kansas to Canada." (Vol. 13, 1948, pp. 438–445).

Thomas, R. P., A. W. Specht, and H. B. Winant. "A Catenary Arrangement of the Soils of Maryland." (Vol. 7, 1942, pp. 487–493).

—————. "Naming Soil Series." (Vol. 10, 1945, pp. 328–334).

Vanderford, H. B., and M. E. Sheffer. "Comparison of Fragipan and Bisequal Profiles of the Gulf Coastal Plain with Soils of Southern Loess Belts." (Vol. 30, 1966, pp. 494–498).

Wascher, W. L., R. P. Hubert, and J. G. Cady. "Loess in the Southern Mississippi Valley: Identification and Distribution of the Loess Sheets." (Vol. 12, 1947, pp. 389–399).

Williams, D. A. "Crops and Soils in Conservation." (Vol. 22, 1958, pp. 350–354).

Winters, E. "Silica Hardpan Development in the Red and Yellow Podzolic Soil Region." (Vol. 7, 1942, pp. 437–440).

Soil Science

Ableiter, K. "Soil Classification in the United States." (Vol. 67, 1949, pp. 183–191).

Cline, M. C. "Basic Principles of Soil Classification." (Vol. 67, 1949, pp. 81–91).

Iver, J. N., and D. H. Frances. "Soil Classification and Soil Maps: Units of Mapping." (Vol. 67, 1949, pp. 163–168).

Mackenhern, R. J., E. P. Whiteside, E. H. Templin, R. F. Chandler, and L. T. Alexander. "Soil Classification and the Genetic Factors of Soil Formation." (Vol. 67, 1949, pp. 93–105).

Moon, J. W., W. S. Ligon, and J. R. Henderson. "Soil Classification and Soil Maps: Original Field Surveys." (Vol. 67, 1949, pp. 169–175).

Orvedal, A. C., M. Baldwin, and A. J. Vessel. "Soil Classification and Soil Maps: Compiled Maps." (Vol. 67, 1949, pp. 177–181).

Reicken, F. F., and G. D. Smith, "Lower Categories of Soil Classification: Family, Series, Type, and Phase." (Vol. 67, 1949, pp. 107–115).

Thorp, J., and G. D. Smith. "High Categories of Soil Classification: Order, Suborder, and Great Soil Groups." (Vol. 67, 1949, pp. 117–126).

Winters, E. "Interpretative Soil Classification: Genetic Groupings." (Vol. 67, 1949, pp. 131–139).

FILMS

Birth of the Soil. 10 minutes, sound, color. Wilmette, Illinois: Encyclopaedia Britannica Films, 1948.

Conservation of Natural Resources. 11 minutes, sound, black and white. Wilmette, Illinois: Encyclopaedia Britannica Films, 1937.

Conserving Our Soil Today. 11 minutes, sound, black and white. Chicago, Illinois: Coronet Instructional Films, 1960.

The Earth: Its Structure. 11 minutes, sound, black and white. Chicago, Illinois: Coronet Instructional Films, 1960.

Erosion. 6 minutes, sound, black and white. Washington, D. C.: Soil Conservation Service, 1948.

Look to the Land. 20 minutes, sound, black and white. Wilmette, Illinois: Encyclopaedia Britannica Films, 1954.

Our Soil Resources. 11 minutes, sound, black and white. Wilmette, Illinois: Encyclopaedia Britannica Films, 1947.

Raindrop and Soil Erosion. 21 minutes, sound, color. Washington, D. C.: U. S. Soil Conservation Service, 1955.

Save the Soil. 11 minutes, sound, black and white. Washington, D. C.: U. S. Department of Agriculture, 1937.

Soil and Water. 8 minutes, sound, black and white. Washington, D. C.: U. S. Soil Conservation Service, U. S. Department of Agriculture, 1962.

Soil and Water Conservation. 8 minutes, sound, black and white. Washington, D. C.: U. S. Soil Conservation Service, 1948.

Soil Explorations. 26 minutes, sound, color. Washington, D. C.: U. S. Army Corps of Engineers, Department of the Army, 1951.

The Story of Soil. 11 minutes, sound, black and white, or color. Chicago, Illinois: Coronet Instructional Films, 1960.

Topsoil. 10 minutes, sound, black and white. Washington, D. C.: U. S. Soil Conservation Service, 1948.

What Is Soil? 10 minutes, sound, black and white. New York, N. Y.: Films, Inc., 1947.

Yours Is the Land. 21 minutes, sound, color. Wilmette, Illinois: Encyclopaedia Britannica Films, 1949.

FILMSTRIPS

The Field Day: Man Cooperates with Nature. New York: Curriculum Films, Inc., 1951.

Soil Resources. New York: Curriculum Films, Inc., 1951. 30 frames, color, 35mm.

Glossary

The purpose of this glossary is to provide meanings or explanations of some of the more technical and unfamiliar terms used in this book. It does not include all such terms, since many of them are defined and discussed when they are introduced. Another group of terms not included in this glossary are names, particularly those that fix a given soil's position in the proper category of a classification. The reader is urged to make use of the two tables in Chapter 4 and of Appendix C to familiarize himself with the use of such terms.

ABC Soil—A soil with a complete profile, including an A, a B, and a C horizon.

Accelerated Erosion—Erosion more rapid than normal geological erosion; activities of man or animals may accelerate erosion.

Acidity—The state of being acid or sour. The water solution of an acid soil turns blue litmus paper red and has a sour taste.

AC Soil—A soil with an incomplete profile, including an A and a C, but not a B horizon. These soils are usually very young, such as those developing from alluvium, from dry, nearly pure sand, or on steep, rocky slopes.

Adsorption—The process of clinging to the surface of a substance. Adhesion of a very thin layer of molecules of a gas or liquids to the surfaces of solid bodies with which they are in contact.

Aeration, Soil—The impregnation of soil with air.

A Horizon—The surface horizon of a soil, lying above the B horizon; sometimes called the topsoil, surface soil, or plow layer. Many soils have lost part of or all their A horizon by erosion.

Alfisol—One of ten terms of the Order (highest) category of the Seventh Approximation Classification. Identified with the order category by the *sol* ending. In many instances these terms serve merely as names, without literal meaning. Most terms in this nomenclature are formed from Greek and Latin roots, with specific syllables to identify them with the proper category. Where the vowel preceding the *sol* ending is *i* the root is Latin, while the vowel *o* indicates a Greek root.

Alkali Soil—The state of the soil when soluble mineral salts are present in the soil. The salts are bases and have a tendency to neutralize acids.

Alluvial Fan—Fan-shaped deposit left by a stream where it emerges from a steep mountain valley upon open level or less steep ground.

Alluvium—Sand, silt, clay, or gravel material deposited by running water.

Arable Land—Land suitable for or cultivated by plowing or tilling.

Aridisols (Arídi-sol, as in *aridity*)—One of ten terms of the Order (highest) category of the Seventh Approximation Classification (see Alfisol). As suggested by the portion *arid,* Aridisols are characteristic of arid climates.

Atom—The smallest part of an element that can take part in a chemical change without being changed itself.

Atomic Mass Number—The total number of protons and neutrons in the nucleus of the atom.

Atomic Number—Number of unit positive charges on the nucleus; equal to the number of protons in the nucleus of an atom.

Atomic Weight—The relative weight of an atom of an element compared with the weight of an atom of carbon 12.

Azonal Soil—Any group of soils without well-developed profiles. These are: Alluvial soils, Lithosols and Dry Sands.

Bar—Mass of sand, gravel, or alluvium deposited on the bed of a stream, sea, or lake.

Bases—Compounds that produce hydroxyl ions in solution and react with and neutralize acids to form salts; bases turn red litmus blue and have a bitter taste and soapy feeling.

Basalt—A dark, granular rock formed from lava.

BC Soil—A soil with an incomplete profile (including a B and a C, but little or no A horizon). Most BC Soils have lost the A horizon by erosion. Sometimes they are called truncated soils, although a truncated soil may have lost all of both A and B.

B Horizon—The soil horizon lying beneath the A horizon; sometimes called the *subsoil*.

Bog Soils—Soils filled with decayed moss and other vegetable matter; they are characterized as being wet, marshy, and spongy.

Brunizem (Prairie soil)—A zonal group of soils having a dark-colored, granular A horizon six or more inches thick which rests on a brownish-colored subsoil, commonly having a blocky structure and usually a higher silicate clay content than the adjoining horizons. The organic content of the surface horizon gradually decreases with depth. The exchange complex contains fewer exchangeable H ions than other cations; these soils are usually developed under grass vegetation in a humid to semihumid climate.

Calcareous Soil—Soil composed of calcium carbonate or containing a high proportion of calcium.

Calcification—The process of depositing insoluble lime salts; becoming stony or calcareous by the secretion of lime salts.

Caliche (ca-*li*-che: *li* as in *liter* and *che* as in *cheese*)—A crust of calcium carbonate ($CaCO_3$) formed on stony soil in arid regions.

Capillary Water—Water held by adhesion and surface-tension forces as a film around particles and in the capillary spaces. It moves in any direction in which capillary tension is greatest.

Carbonation—Chemical weathering in which minerals are altered to carbonates by carbonic acid.

Catch Crop—A crop seeded with one of the regular crops in a rotation or between the growing periods of two regular crops for the purpose of adding organic matter or nitrogen to the soil, or for producing hay or other crops of economic value.

Catena—A group of soils within a specific soil zone developed from similar parent materials, but with different soil characteristics owing to differences in relief or drainage.

Chernozem Soils—A zonal group of soils having a deep, dark-colored to nearly black surface horizon, rich in organic matter, which grades below into lighter-colored soil and finally into a layer of lime accumulation; developed under tall and mixed grasses in a temperate to cool subhumid climate. The term is derived from the Russian words meaning "black earth."

C Horizon—The soil horizon lying beneath the B horizon, and from which the A and B horizons are formed. It is sometimes called the parent material.

Clay—The small mineral soil particles less than 0.002 millimeters in diameter.

Clayey—Adjective of clay.

Claypan—A dense and heavy soil horizon underlying the upper part of the soil; hard when dry and plastic when wet. This kind of horizon is the most prominent feature of many Planosols.

Coastal Plain—Any plain that has its margin on the shore of a large body of water, particularly the sea.

Cohesion—State of cohering or cleaving together. Sticking together.

Colloidal Condition (Colloid)—A state in which a substance composed of extremely small clusters of molecules can remain suspended in liquids, solids, or gases for long periods of time. Colloidal material is more finely divided than a suspension, less finely divided than a true solution.

Colluvium—Mixed deposits of rock fragments and soil material at the base of comparatively steep slopes, accumulated through slides, "mud flows," and local wash.

Combining Number—The number of atoms that combine with or replace an atom of hydrogen. The combining number of oxygen is 2.

Compound—A substance formed by the chemical combination of definite proportions of two or more elements.

Concretion—A solid, rounded rock formed inside a rock of a different kind that is produced by deposition from aqueous solution in the rock.

Conglomerate—A sedimentary rock composed of gravel, or rounded pebbles, held together by hardened clay, silica, or other similar materials.

Conservator, Resource—A layman (not a professional) who practices resource conservation and accepts a personal responsibility for improving the use of resources as an obligation of citizenship.

Contour Plowing—Plowing around a hill at uniform levels rather than up and down it.

Crust—Layer of granite or basalt rock forming the outer surface of earth's bedrock. The exterior, relatively cool part of the earth.

Decomposition—The breaking down of a chemical compound into simpler compounds or into elements.

Deflation—The erosive action of wind in blowing soil and rock particles from one place to another.

Degradation—Process of degrading, degenerating, or deteriorating.

Delta—A level fan-shaped deposit formed at the mouth of a stream.

Deposition—The laying down of rock or soil materials.

Depression—Low place or hollow surrounded by higher ground.

D Horizon—A horizon recognized in some soils but usually not identified. It underlies the C horizon, or weathered parent material. Usually it consists of undisturbed, unaltered, crustal rock, gravel, or unconsolidated sand.

Displacement—A chemical change in which an element takes the place of another element in a compound, setting the other element free.

Divide—Ridge on high ground that separates drainage basins of streams.

Double Replacement—A chemical reaction between two compounds to form two new compounds by exchanging ions.

Drumlin—A long, low mound of glacial till, rounded at one end and pointed at the other. Sometimes refers to a smooth, oval, or elongated hill or ridge of glacial till.

Electron—An elementary particle having a negative charge and a mass equal to 1/1836 the mass of a proton. Electrons move around the nucleus of the atom, which has a positive charge, or move in a stream from one atom to another as an electric charge.

Element—A substance which cannot be broken down into simpler substances by ordinary chemical means.

Eluviation—The movement of soil material from one place to another within the soil, in solution or in suspension, when there is an excess of rainfall over evaporation. Horizons that have lost material through eluviation are referred to as *eluvial* and those that have received material as *illuvial*.

Emergence—Process by which part of a sea or lake floor becomes dry land.

Entisol—One of ten terms of the Order (highest) category of the Seventh Approximation Classification (see *Alfisol*).

Eolian Material—Material deposited by wind.

Eon—The largest division of geologic time.

Equation—A representation of a chemical change in symbol form. $2H_2 + O_2 \rightarrow 2H_2O$ is an equation that represents the union of hydrogen and oxygen to form water.

Erosion—The removal of soil and rock fragments by moving forces such as surface water, ground water, wind, waves, or glaciers.

Estuary—The wide mouth of a river that flows into the sea and into which the tide flows.

Fall Line—An area marked by rapids and waterfalls, where rivers descend abruptly from an upland to a lowland.

Family, Soil—A category in soil classification between soil series and great soil group or subgroup; it is composed of one or more distinct soil series.

Fertilizer—Material added to the soil to promote the growth of plants. Meager crops will be produced when there is a lack of fertilizer if the soil does not possess needed food materials.

Flash Flood—A sudden flood caused by heavy rains or cloudbursts in the surrounding area.

Flocculation—The coagulation of finely divided particles into particles of greater mass.

Floodplain—A plain or stream valley that borders a stream and is covered by its waters during flood stage.

Formula—A group of symbols and figures representing an element or a compound and showing its composition. The formula for carbon dioxide is CO_2.

Fossil—Remains of ancient animals and plants found naturally preserved in the earth's crust.

Friable—Easily crumbled into small pieces or pulverized.

Gas—A substance that has no definite size or shape, hence is capable of expanding indefinitely.

Genesis, Soil—Mode of origin of the soil, referring particularly to the factors involved in the development of the solum from the unconsolidated parent material.

Geode—A nodule of stone usually lined or filled with crystals or mineral matter formed by deposition in a rock cavity.

Geology—The science of the origin, history, structure, and the chemical and physical nature of the earth's crust.

Glacial Drift—A general term for the rock debris that has been transported by glaciers and is deposited directly or indirectly by the melting of a glacier.

Glacial Lake—A lake basin which has been formed and altered by glacial action.

Gley, Gley Soils—Mineral soils which have mottled or irregular coloration due to alternate dry and wet conditions.

Gneiss—A banded or foliated metamorphic rock corresponding in composition to granite.

Gradient—The degree of slope or of change in the level of roads and streams.

Granite—A common igneous rock containing feldspar and quartz.

Great Group—Category term in the Seventh Approximation Classification, 1960, corresponding to the Great Soil Group category of the earlier classification.

Great Soil Group—A broad group of soils with fundamental internal characteristics in common. (This term represented a category in the American Classification prior to 1960. See Table 1, Chapter 4.)

Ground Water—Subsurface water that accumulates in the zone of saturation. It forms the source of wells and springs.

Grumusol—A group of calcareous grassland soils, found largely in Texas, Alabama, Mississippi, and California; it was identified and named about 1956.

Gully—A miniature valley formed on a hillside by heavy rains.

Gullying—A type of erosion in which deep channels form as water flows down a slope.

Hardness—The resistance of a substance to scratching.

Hardpan—A hardened, or cemented soil horizon. The soil may have any texture and is compacted or cemented by iron oxide, organic material, silica, or other substances.

Hard water—Water which contains dissolved compounds of calcium, magnesium, or iron.

Headland—A projecting point of land along a coast.

Herbicide—A chemical applied to the soil for the purpose of controlling the growth of weeds and other vegetation on land planted in crops.

Histosol—One of ten terms of the Order (highest) category of the Seventh Approximation Classification (see *Alfisol*). As suggested by the *hist,* meaning tissue, Histosols are organic soils.

Humus—The organic material in the soil produced by the decomposition of plant and animal materials.

Hydration—The chemical combination of substances with water.

Hydrolysis—A chemical process in which the ions of a salt react with the ions of water to produce a solution that is either acidic or basic.

Hydroxide—A compound containing the radical ($\overline{OH}$). Soluble metallic hydroxides are bases.

Hydroxyl—The chemical radical ($\overline{OH}$) made up of a single atom of oxygen and one of hydrogen and having a single negative charge.

Hygroscopic Water—Atmospheric water held on the surface of particles by forces of adhesion.

Igneous Rocks—Rocks formed by the cooling and solidification of magma.

Illuviation—See *eluviation*.

Immature Soil—A young or imperfectly developed soil that has not yet come into equilibrium with its environment.

Impervious Rocks—Rocks through which water cannot pass because the pore spaces are so small. Shale is an example.

Inceptisol (Incépti-sol as in *inception*)—One of ten terms of the Order (highest) category of the Seventh Approximation Classification (see *Alfisol*). As suggested by the portion *incept,* Inceptisols are young soils, just beginning their formation.

Inorganic—Refers to compounds that do not contain carbon, except carbonates and cyanides. Ordinary table salt is an inorganic compound.

Intrazonal Soil—Any of the great groups of soils with more or less well-developed soil characteristics that reflect the dominating influence of some local factor of relief, parent material, or age over the normal effects of the climate and vegetation.

Intrusive Igneous Rocks—Molten rock which hardened in cracks and openings in other rock layers.

Intrusive Rocks—Rocks formed when the molten magma solidified among other rocks. The rocks, when molten, had been forced between consolidated rock, including layers of sedimentary rocks.

Ion—A particle with an electric charge, formed by the gain or loss of one or more electrons. (An atom or radical that carries an electric charge.)

Irrigation—Artificial watering of farm land by canals, ditches, sprinklers, underground conduits, or flooding to supply growing crops with moisture.

Isotopes—Atoms of the same element which have the same atomic numbers but different atomic weights.

Lacustrine Plains—Lake plains which were formed by the emergence of a lake floor by either uplift or drainage.

Land Capability Classes—A system of classifying the land, usually farm by farm, to express the most desirable and practical use to be made of each land parcel.

Laterization—A geologic process most common in tropical climates that produces a laterite, a bright red substance high in iron oxide.

Latozation—A soil-forming process, closely related to the geologic laterization process that produces lateritic soils.

Lava—Liquid (molten) rock material that flows out on the surface of the earth.

Leaching—The removal of minerals from soil and rock by solution as water seeps down from the surface.

Levee—A bank that confines a stream to its channel. It may be natural or artificial.

Lithosol (Skeletal Soils)—A great soil group having no clearly expressed soil morphology and consisting of a freshly and imperfectly weathered mass of rock fragments; largely confined to steeply sloping land. (*Lithol* is from the Greek for rock.)

Lithosphere—Solid part of the earth (crust, mantle, and core).

Litmus—A blue-violet dye extracted from plants called lichens. Litmus is used as an acid-base indicator.

Loëss—An extensive deposit of windblown silt.

Magma—Huge masses of molten (hot liquid) rock beneath the earth's surface. Igneous rocks are formed from magma.

Mantle—Zone of rock extending from the crust downward to about 1800 miles.

Mantle Rock—A layer of loose rock that covers the bedrock.

Marl—An earthy, crumbling deposit consisting chiefly of clay mixed with calcium carbonate. It is used as fertilizer for soils deficient in lime.

Mature Soil—A soil with well-developed characteristics produced by the natural processes of soil formation and in equilibrium with its environment.

Meander—One of a series of regular bends in a stream.

Metamorphic Rocks—Those rocks formed by the effect of heat, pressure, and chemical action on other rocks.

Mineral—An inorganic substance of definite composition found in the earth's crust.

Molecule—The smallest unit quantity of matter that can exist by itself and still retain the composition and properties of a larger amount of the substance.

Mollisol—One of ten terms of the Order (highest) category of the Seventh Approximation Classification (see *Alfisol*). As suggested by the portion *moll,* Mollisols are "soft" soils, usually having a relatively thick, dark surface layer, that is high in organic content.

Moraine—A ridge, mound, or undulating plain of boulders, gravel, sand, and clay deposited by a glacier.

Morphology—The scientific study of the structure and forms of land in relation to the development of topographic features produced by erosion.

Mottled—Marked or streaked with spots of different colors. Mottled soils indicate poor drainage.

Mouth—Exit or point of discharge of a stream into another stream, lake, or sea.

Muck—Fairly well composed organic soil material, relatively high in mineral content (40 to 50%), dark in color, and accumulated under conditions of very poor drainage.

Mulch—Substances such as straw or moss spread upon the ground to protect the roots of plants from heat, cold, or drought.

Natural Levees—Banks along a stream built up by sediments deposited during flood tide.

Neutralization—A complete action between an acid and a base so that the products have the characteristics of neither.

Neutron—A sub-atomic particle with a mass slightly greater than that of the proton.

Nitrogen Cycle—The complex series of actions by which nitrogen is successively a part of the air, the soil, plants, animals, and finally the air again.

Nodule—A concretion that contains minerals.

Nomenclature, Soil—A system of names used in soil science.

Nucleus—The positively charged, dense central portion of the atom, consisting of neutrons and protons.

Nutrient, Plant—The elements or groups of elements taken in by the plant, essential to its growth, and used by it in the growth of its tissue and in the development of seed.

Offshore Bar—A sand bar more or less parallel to the mainland or shoreline.

Order Category—The highest category of soil classification in both the new Seventh Approximation Classification and the older American Classification.

Organic—Pertaining to or derived from living organisms; also designating compounds containing carbon.

Organic Soil—A general term used in reference to any soil, the solid part of which is predominantly organic matter.

Outcrop—Upturned edges of rock layers exposed at the earth's surface.

Outwash Plains—Plains formed by the deposition of materials by streams or sheet flow from a melting glacier.

Oxidation—The chemical union of oxygen with other substances.

Oxide—A binary compound containing oxygen and another element.

Oxisol—One of ten terms of the Order (highest) category of the Seventh Approximation Classification (see *Alfisol*), As suggested by the syllable ox, Oxisols are highly oxidized as a result of advanced weathering.

Parent Material, Soil—The unconsolidated mass from which the soil profile develops. The C Horizon is referred to as parent material.

Peat—Dark brown decomposed plant material. It represents the first stage in coal formation.

Pedalfer (Pe-dal-fer)—A soil in which there has been a shifting of aluminum and iron oxide downward in the soil profile but with no horizon of carbonate. Roughly equivalent to soils of the humid regions. Term derived from soil (*ped*), aluminum (*al*), and iron (*fer*).

Pedocal (Ped-o-cal)—A soil with a horizon of accumulated carbonates, the lime zone, in the lower part of the solum or just beneath it. Roughly equivalent to soils of arid and semiarid regions. Term derived from soil (*ped*) and lime (*cal*).

Pedologic Process—Processes of soil formation.

Pedologic Time—The time required by nature to produce a soil through normal pedologic processes. Not a widely used term in soil literature but it provides a convenience for distinguishing soil forming time from geologic time, historical time, and other time references.

Pedology—A science that treats the study of soils.

Pedon—The smallest volume that can be classed as a soil. Its surface area ranges from 1 to 10 square meters. Introduced as a soil term in the Seventh Approximation Classification, 1960.

Peneplain—A low, rolling, nearly level surface formed by weathering and erosion.

Percolation—The process of oozing through some porous substance.

Periodic Table—A table of elements giving all basic information about the elements.

Plain—Region of horizontal rock structure and low relief due to low elevation.

Plateau—Region of horizontal rock structure and high relief due to higher elevation.

Podzol—A group of zonal soils that develops in a cold, moist climate under coniferous and deciduous forests. It has a thin organic mineral layer above a characteristic gray, leached layer.

Podzolization—The process of soil development in humid regions involving the leaching of the upper layers of soil and accumulation of material in the lower layers with the resultant development of characteristic horizons.

Porosity—The state of being porous enough for liquids to pass through.

Precipitation—Moisture that has condensed and come out of the air in the form of fog, dew, frost, rain, hail, snow, or sleet. Also the separating out of a substance from a solution because of a chemical or physical change. The formation of an insoluble compound in a solution.

Proton—An elementary part of the atomic nucleus, having one unit of positive charge.

Pumice—Surface lava which looks like foam and hardens into a spongy rock.

Radical—A group of atoms which act as a single atom in a reaction. OH is an example of a radical.

Raw Humus—Undecomposed organic matter in soil.

Reaction—A chemical change. Some reactions are promoted by catalysts.

Recommended Farming Practices—A term used in agricultural literature to denote desirable farming procedure as determined by agricultural experts and soil scientists in cooperation with farmers of a given area.

Red-Yellow Podzolic Soil—A great soil group in the American Classification prior to 1960. The dominant zonal group of soils in Southeastern United States and in other world regions with similar climates. Normally moderately to strongly acid, low in organic matter, and strongly leached. The better-drained upland members are red, the poorly drained members tend to be yellow or gray.

Regolith—Mantle rock.

Regosol—An azonal soil consisting chiefly of soft consolidated material such as sand or volcanic ash. It has little evidence of pedological development.

Relief—The irregularity in elevation of parts of the earth's surface.

Rendzina—A group of dark-colored soils usually developing under tall grass or vegetation from highly calcareous parent material. Found in several soil zones, although they are typical of rainy, middle-latitude climate where the high lime content of the parent material is not leached to produce an acid reaction.

Residual Soil—Soil formed from decomposed bedrock material remaining on the surface in the area where it was formed.

Ridge—A narrow, elongated crest of a hill or mountain; a range of hills or mountains.

Rock-forming Minerals—Common minerals which make up large percentages of the rocks of the earth's crust.

Saline Soil—A soil containing an excess of soluble salts and not very highly alkaline; sometimes called salty soils.

Salts—Compounds which contain a metal and a non-metal and are neutral to litmus paper. There is a positive ion other than hydrogen and a negative ion other than a hydroxyl.

Sedimentary Rocks—Rocks formed in layers from materials deposited by water, wind, ice, or other agents.

Series, Soil—A group of soils having similar profiles except for the texture of the surface soil, and developed from a particular type of parent material. A series may include two or more soil types differing from one another in the texture of the surface soil.

Seventh Approximation, Classification—A soil classification of world scope developed in the United States and introduced in 1960. It differs widely from other classifications. It introduces a new nomenclature of technical terms derived from Greek and Latin. It is highly complex, but very functional if understood and applied properly. It became the official classification for mapping American soils in 1965.

Silt—Soil particles intermediate in size between clay particles and sand grains, which are suspended in or deposited by water.

Sink—A depression in the earth's surface formed by the collapse of the roof of an underground cavern. Soluble bedrock such as limestone often forms sinks.

Sod Farming—A term sometimes used to describe a system of cropping where planting in unprepared sod or other plant cover is followed by application of chemicals including herbicides, and where no tillage is involved.

Soil—The natural medium for the growth of land plants on the earth's surface. A natural body on the surface of the earth in which plants grow; composed of mineral and organic materials.

Soil Conservationist—A technically trained person who prescribes, administers, or supervises the various conservation techniques and programs of the Soil Conservation Service.

Soil Conservation Service—A service of the United States government devoted to improving agricultural practices through cooperation with other government agencies and with individual farmers.

Soil Horizon—A layer comprising a distinct part of a soil. A combination of the soil horizons—usually three—comprises the soil profile.

Solum—The layer of soil which lies above the parent material in which the natural processes of soil formation take place.

Solute—A substance dissolved in a solvent.

Solution—A homogeneous mixture of solvent and solute. Water and sugar shaken together form a solution.

Solvent—A liquid that dissolves a substance. Water is a solvent.

Spodosol—One of ten terms of the Order (highest) category of the Seventh Approximation Classification (see *Alfisol*). As suggested by the syllable *spod,* a Greek term meaning wood ashes, Spodosols include Podzols and other ash-colored soils.

Steppes—Regions of semi-arid grasslands in middle latitudes. They often border tropical deserts.

Structure, Soil—The morphological aggregates in which the individual soil particles are arranged.

Subgroup—The fourth category in the Seventh Approximation Classification preceded by the order, the suborder, and the Great Group.

Suborder—The second highest category in both the new Seventh Approximation and the older American Classification.

Subsoil—A layer of weathered rock lying below the topsoil.

Terrace—Geologically, a flat or undulating plain, commonly rather narrow and usually with a steep front bordering a stream. Some streams are bordered by a series of terraces, one above the other, each terrace representing a former floodplain. Marine terraces on some coastal margins resemble river terraces and have resulted from periodic uplifting of the coastal margins. Man-made terraces of many kinds serve to control erosion and otherwise improve agriculture.

Texture, Soil—The relative proportion of fine and coarse particles in the soil.

Till—Unstratified glacial deposits which are composed chiefly of clay and stones.

Tilth—The physical condition of a soil in respect to its fitness for the growth of a specified plant or a plant sequence.

Topographic Map—A map that shows surface features of a region by the use of contour lines and symbols.

Topography—The physical features of a region.

Topsoil—A vague, general term sometimes applied to the surface portion of the soil, including the average plow-depth (plow-layer) or the A Horizon, whichever is thicker.

Trace Element—A chemical element (as zinc, boron, or iodine) found combined in minute quantities in plant or animal tissues and considered essential in the physiological process of most plants and animals. It is often called a microelement or micronutrient.

Transported Mantle Rock—Mantle rock that has been moved from its place of origin.

Travertine—Ground water deposit composed chiefly of calcite.

Tundra—A region which is continuously cold and damp. It is usually characterized by the growth of mosses, lichens, ferns, and small shrubs.

Ultisol—One of ten terms of the Order (highest) category of the Seventh Approximation Classification (see *Alfisol*). As suggested by the syllable *ult,* meaning last, Ultisols are very old soils of humid climates.

Unconformity—A discontinuity in rock sequences, indicating an interruption of sedimentation.

Unstratified—Rocks that are massive and not formed in beds or strata.

Valence—The combining power of an element or radical, measured by the number of hydrogen atoms with which the radical or one atom of the element will combine or which it will replace in a chemical reaction.

Valley—A depression in the land surface which is elongated and usually has a stream.

Vertisol—One of ten terms of the Order (highest) category of the Seventh Approximation Classification (see *Alfisol*). As suggested by the syllable *vert* (to turn, invert, or break), Vertisols tend to crack badly on drying out (and to overturn and slip in the process).

Virgin Soil—Soil which has not been plowed or tilled.

Waterlogged Soil—Soil in which the water table has risen high enough to expel normal soil gases and to interfere with plant growth or cultivation.

Watershed—The entire area drained by a stream.

Water Table—The surface below which the ground is saturated with water.

Weathering—The natural disintegration and decomposition of rocks and minerals.

Zonal Soils—A major soil group often classified as a category of the highest rank and generally covering a wide geographic region or zone and embracing soils that are well developed from the parent material by the normal soil-forming action of climate and living organisms.

PERIODIC CLASSIFICATION OF THE ELEMENTS*

IA	IIA	IIIB	IVB	VB	VIB	VIIB		VIIIB		IB	IIB	IIIA	IVA	VA	VIA	VIIA	0
1 H 1.00797																	2 He 4.0026
3 Li 6.939	4 Be 9.0122											5 B 10.811	6 C 12.01115	7 N 14.0067	8 O 15.9994	9 F 18.9984	10 Ne 20.183
11 Na 22.9898	12 Mg 24.312											13 Al 26.9815	14 Si 28.086	15 P 30.9738	16 S 32.064	17 Cl 35.453	18 Ar 39.948
19 K 39.102	20 Ca 40.08	21 Sc 44.956	22 Ti 47.90	23 V 50.942	24 Cr 51.996	25 Mn 54.9380		26 Fe 55.847 27 Co 58.9332 28 Ni 58.71		29 Cu 63.54	30 Zn 65.37	31 Ga 69.72	32 Ge 72.59	33 As 74.9216	34 Se 78.96	35 Br 79.909	36 Kr 83.80
37 Rb 85.47	38 Sr 87.62	39 Y 88.905	40 Zr 91.22	41 Nb 92.906	42 Mo 95.94	43 Tc (97)		44 Ru 101.07 45 Rh 102.905 46 Pd 106.4		47 Ag 107.870	48 Cd 112.40	49 In 114.82	50 Sn 118.69	51 Sb 121.75	52 Te 127.60	53 I 126.9044	54 Xe 131.30
55 Cs 132.905	56 Ba 137.34	57 La 138.91	72 Hf 178.49	73 Ta 180.948	74 W 183.85	75 Re 186.2		76 Os 190.2 77 Ir 192.2 78 Pt 195.09		79 Au 196.967	80 Hg 200.59	81 Tl 204.37	82 Pb 207.19	83 Bi 208.980	84 Po (209)	85 At (210)	86 Rn (222)
87 Fr (223)	88 Ra (226)	89 Ac (227)															

58 Ce 140.12	59 Pr 140.907	60 Nd 144.24	61 Pm (145)	62 Sm 150.35	63 Eu 151.96	64 Gd 157.25	65 Tb 158.924	66 Dy 162.50	67 Ho 164.930	68 Er 167.26	69 Tm 168.934	70 Yb 173.04	71 Lu 174.97
90 Th 232.038	91 Pa (231)	92 U 238.03	93 Np (237)	94 Pu (244)	95 Am (243)	96 Cm (247)	97 Bk (247)	98 Cf (251)	99 Es (254)	100 Fm (253)	101 Md (256)	102 No (253)	103 Lw (257)

*Mass numbers of isotopes with longest half lives given in parentheses.

Index